QUANTUM ELECTRON THEORY OF AMORPHOUS CONDUCTORS

KVANTOVO-ELEKTRONNAYA TEORIYA AMORFNYKH PROVODNIKOV

КВАНТОВО-ЭЛЕКТРОННАЯ ТЕОРИЯ АМОРФНЫХ ПРОВОДНИКОВ

QUANTUM ELECTRON THEORY OF AMORPHOUS CONDUCTORS

by

Aleksandr Ivanovich Gubanov

Translated from Russian by
A. Tybulewicz
Editor, *Physics Abstracts* and *Current Papers in Physics*

CONSULTANTS BUREAU
NEW YORK
1965

First Printing — April 1965

Second Printing — August 1966

The original Russian text, published for the Physical-Technical Institute of the Academy of Sciences of the USSR by the Academy of Sciences Press in Moscow in 1963, was revised and updated by the author for the English edition.

Александр Иванович Губанов

КВАНТОВО-ЭЛЕКТРОННАЯ ТЕОРИЯ
АМОРФНЫХ ПРОВОДНИКОВ

Library of Congress Catalog Card Number 65-10526

ISBN-13: 978-1-4684-0669-6   e-ISBN-13: 978-1-4684-0667-2
DOI: 10.1007/978-1-4684-0667-2

PREFACE

The electron theory of solids has attracted great attention in recent years, mainly because of the numerous practical applications of semiconductors. However, all the reviews and monographs on this subject deal only with crystalline conductors. At present, mainly in the Soviet Union, experimental and theoretical investigations have been extended to liquid and solid amorphous conductors, and in particular to such semiconductors. However, all the work published so far in this field is in the form of separate papers scattered throughout various journals, and there has as yet been no Soviet or foreign review of the theoretical work on amorphous semiconductors, in spite of the increasing interest in them. The investigation of liquid and amorphous semiconductors is of great practical importance, first, because all the solid semiconductors are usually prepared from the liquid phase and it is important to know the electrical and other properties of this phase; secondly, amorphous semiconductors are beginning to be used in industry, for example, amorphous Sb_2S_3 films in vidicon tubes. In some cases, especially in optical instruments, amorphous semiconductors have advantages compared with crystals. Theoretical studies of amorphous semiconductors should help in these practical applications.

The present monograph is the first attempt to present systematically the quantum electron theory of amorphous conductors. The most interesting—in the author's view—theoretical papers on this subject, published in journals are reviewed and critically compared.

The book begins with a review of experimental data on liquid and amorphous conductors: the electrical conductivity of liquid metals, changes in the properties of semiconductors on melting, and investigations of glassy semiconductors (Chap. I). Then the fundamentals of the quantum electron theory of solids are given in Chap. II. This chaper is included because the theory of amorphous conductors starts from solid-state theory and uses to a considerable extent the same calculation methods. Therefore it was essential to give the basis of these methods, in particular the widely used one-electron approximation. On the other hand, over the greater part of the book, the electron spectrum and wave functions in crystals are assumed to be known. Chapter II gives information on the methods of calculating these quantities.

Chapter III shows the similarities and differences between the structures of liquid and crystalline substances. This chaper presents experimental justification for the model of a liquid which is used as the basis of the theory of liquid and amorphous conductors developed in Chaps. IV–IX.

Chapters IV–IX give the various theories of liquid and amorphous electronic conductors. Chapter IV deduces, using one-dimensional models of liquids, the band structure of the electron energy spectrum. In Chap. V, this theory is extended to the three-dimensional model. The characteristic features of electron scattering in amorphous substances are discussed in Chaps. VI and VII. The mean free path of electrons and the temperature dependence of various galvano-thermomagnetic coefficients of an amorphous substance are calculated in Chap. VIII. The local and impurity levels in amorphous semiconductors are dealt with in Chap. IX, which also offers an explanation of the absence of impurity conduction in glassy semiconductors.

Chapters X and XI are devoted to a theory of liquid metals, the approach to the solution of the problem being essentially different from that in Chaps. IV–IX, which covered all liquid conductors, including metals.

The last (short) chapter – XII – deals with the theory of amorphous ferromagnets, which is not based on calculations given in previous chapters.

The problems of ionic conduction in liquid electrolytes, glasses, and other amorphous substances have been dealt with in sufficient detail in the literature and are not discussed in the present monograph.

The book does not deal with the recent numerous theoretical papers on various types of defective crystal structure: disordered alloys, strongly doped semiconductors, etc. These papers should be the subject of a separate monograph because, in spite of some similarity of the calculation methods, the basis of the theory is different from that for amorphous conductors where the emphasis is on the absence of any, even approximate, long-range order in the distribution of the atoms.

INTRODUCTION

The application of quantum mechanics to solids has resulted in the very
successful explanation of their electrical properties. The high electrical conductivity of metals and its temperature dependence became understood, then
the differences between the electrical conductivities (and their temperature
dependences) of metals, insulators, and semiconductors were explained. The
most remarkable achievements of the quantum-mechanical theory of solids
occurred in the development of semiconductor physics. Electron and hole conduction, the considerable influence of negligibly small amounts of impurities,
the different effective masses of carriers, and many similar phenomena, which
were quite incomprehensible within the framework of the classical electron
theory, were explained in a natural way by quantum mechanics.

The modern theory of solids is based on the "band theory," which is used
directly or indirectly in the great majority of theoretical papers on metals and
semiconductors.

The essence of the orthodox band theory is as follows. The one-electron approximation is assumed, i.e., the motion of a single electron is considered in a self-consistent field produced by the lattice ions and the aver-
aged-out charge distribution of all the other electrons. This field has the periodicity of the crystal lattice and the mathematical problem reduces to the
solution of the single-particle Schrödinger equation with a periodic potential.

Irrespective of the actual form of the potential, and using only its periodicity, the essential properties of the Schrödinger equation solution are obtained in a mathematically rigorous way. These properties are: the energy
spectrum in the form of several continuous allowed bands, separated by forbidden bands; wave functions in the form of plane waves, modulated at the
lattice period (Bloch functions) and describing free motion of an electron within
a crystal, the effective mass of the electron being different from its mass
in vacuum. Thus, the periodically distributed atoms or ions do not prevent
translational motion of an electron, scattering occurring only on departure
from periodicity.

In accordance with the Pauli principle, each allowed band has a limited number of electron vacancies, which, in the simplest case, is equal to double the number of atoms in a crystal. All electrons in a crystal are distributed over such levels, filling the bands beginning with the lowest one. If one of the bands is partly filled, we have a metal with good electrical conductivity because the periodically distributed atoms do not impede electron motion in an external field; electrons are scattered mainly by the thermal motion of atoms, which causes the electrical resistivity of a metal to rise with temperature. If several bands are completely filled and the nearest empty band is separated by a gap, conduction is prevented by the absence of free levels in the filled bands and we have an insulator or a semiconductor. Conduction is possible only when electrons are transferred thermally from a filled to a free band or to an impurity level, or from a filled impurity level to a free band. Now the electrical conductivity increases with increase of temperature and with increase of the impurity concentration; moreover, it may be of n-type (electron) or p-type (hole conduction). The properties of semiconductors are very sensitive to the energy spectrum structure and the existence of semiconductors and crystalline insulators depends on the presence of a forbidden gap.

We have used the language of the one-electron theory to discuss the problem of electrical conduction, but the results are essentially the same in the many-electron theory of crystals.

The quantum-mechanical explanation of the electrical properties of metals and semiconductors is based on the periodicity of the potential and the presence of long-range order in the distribution of atoms. This contradicts the old established observation that the electrical conductivity of a metal does not change greatly on melting and that it retains its metallic nature. Liquid mercury is a typical metal in its electrical and optical properties.

This discrepancy between theory and experiment could have been accepted if it were true only of metals, because the properties of metals are not very sensitive to the structure of the electron energy spectrum and the most important result of the band theory of metals is the prediction of a high electrical conductivity. However, recent investigations of changes in the properties of semiconductors on melting showed that in some cases their electrical conductivity and other properties change little on melting and retain their semiconducting nature. The question is why semiconductors, which are so sensitive to changes in their band structure, do not change their properties upon the disappearance of the periodicity of the atomic distribution, which is responsible for the presence of allowed and forbidden bands in the energy spectrum.

The problem facing theoreticians was to prove that when the strict periodicity or long-range order in the atomic distribution disappears, i.e., in a liquid or in an amorphous substance, the following features are retained:

1) forbidden bands in the energy spectrum;

2) allowed energy levels, forming continuous or quasicontinuous bands;

3) electron behavior is described, at least in the first approximation, by certain waves propagated in the substance, i.e., electrons are quasifree, as in crystals.

After proving these properties, one can develop the quantum electron theory of amorphous substances using the standard solid-state theory as a model, but allowing for the features connected with the absence of long-range order, such as, for example, additional electron scattering characteristic of amorphous substances.

First, quantum-mechanical calculations of solids were carried out using one-dimensional models of crystals. In spite of the arbitrary nature of the one-dimensional model, it gives many interesting results which follow from simple calculations, and these results can be extended to three-dimensional crystals. Thus, the simplest one-dimensional model of Kronig and Penney yielded energy bands, explicit expressions for modulated direct waves, Tamm surface levels, etc. Naturally enough, the theory of amorphous substances starts with a one-dimensional model.

However, the calculations relating to a one-dimensional model of an amorphous substance, i.e., assuming a chain of nonuniformly distributed atoms or cells, are considerably more difficult than those relating to a one-dimensional model of a crystal. We cannot use the periodicity condition and thus reduce all the equations to one equation for a single cell. The problem can be solved in various ways.

One method considers a chain of a finite, but sufficiently large, number of cells and solves the system of equations for all the cells. Such a solution can be obtained only numerically, which was done by means of electronic computers. The results of such calculations are given at the end of Chap. IV.

Another method uses the assumption of quasicrystallinity of an amorphous substance, i.e., that in such a substance the distances between neighboring atoms fluctuate at random but on the average they do not differ greatly from some mean value. Then we can introduce such a coordinate scale in which the chain potential is approximately periodic. Such a calculation takes up the major part of Chap. IV. The quasicrystalline model is justified by the experimental studies of the structure of liquid and amorphous substances described in Chap. III.   The quasicrystalline approach can be extended to the three-dimensional model of a liquid, as shown in Chap. V.

The third method of dealing with a liquid, elucidated in Chaps. X and XI, is applicable to three-dimensional models and is based on the method of weakly bound electrons. The scattering of plane electron waves by individual atoms is considered, allowing for the correlation between the atoms with some order in their distribution. Such calculations are applicable only to liquid metals which are mostly univalent.

Theoretical studies show that the energy spectrum and wave functions of electrons in a liquid have, in the first approximation, the same nature as in a crystal with similar short-range order. The energy spectrum structure and other characteristics are thus determined not by the long-range but by the short-range order in the distribution of atoms. This point of view was repeatedly put forward by Academician A. F. Ioffe and has now been confirmed by experimental studies and theoretical calculation.

CONTENTS

Chapter I

Review of Experimental Data on Liquid and Amorphous Conductors

Chapter II

Fundamentals of the Electron Theory of Solids

Chapter III

Structure of Amorphous Substances

Chapter IV

Band Theory for the One-Dimensional Model of a Liquid

Chapter V

Band Theory for the Three-Dimensional Model of a Liquid [1]

Chapter VI

Scattering of Electrons in Liquids by Departures from Long-Range Order and by Defects

Chapter VII

Scattering of Electrons on Thermal Vibrations in a Liquid

Chapter VIII

Electrical Conductivity, Thermal Conductivity, Thermoelectric Power, Hall Coefficient, and Nernst Coefficient of Amorphous Substances Exhibiting Electronic Conduction

Chapter IX

Local and Impurity Levels in Amorphous Semiconductors

Chapter X

Electron Structure of Liquid Metals

Chapter XI

Theory of the Electrical Conductivity of Liquid Metals

Chapter XII

Quasi-Classical Theory of Amorphous Ferromagnets

* * *

Chapter I

REVIEW OF EXPERIMENTAL DATA ON LIQUID AND AMORPHOUS CONDUCTORS

§ 1.   Electrical Conductivity of Liquid Metals

The electrical conductivity of the majority of metals drops by a factor of 1.5-2 on melting, but there are some anomalous metals whose electrical conductivity is higher in the liquid than in the solid state. Table 1 lists the values of the ratio of the electrical conductivities of several metals in the solid states at the melting point. These values—listed in order of decreasing σ_s/σ_l—are taken from Regel' [1], and Roll and Motz [2], who critically reviewed previous work [3-16]. Table 1 also contains the coordination numbers z of metals in the solid and liquid states, as well as their percentage change of volume on melting.

Table 1 shows that the change in the electrical conductivity of metals on fusion is relatively small: with few exceptions the reduction is not greater than by a factor of 2.5. The electrical conductivity differs much more than that from metal to metal in the solid state; for example, the electrical conductivity of potassium at room temperature is 200 times higher than that of bismuth, and even in similar metals, such as silver and platinum, the electrical conductivity differs by a factor of nearly 7. Thus, without doubt, we can state that the mechanism of electrical conduction in solid and liquid metals is the same: the metallic nature of conduction is retained after fusion.

The first attempt to present a systematic account of the change in the electrical conductivity of metals on melting was given by Perlitz [17]. He assumed that in the melt the atoms are distributed at random so that the change in the electrical properties on melting is determined by the structure of the metal in the solid state. On the basis of the then available experimental data, Perlitz established that the resistivity of close-packed metals is doubled on fusion, that of metals with a cubic structure increases by a factor of 1.5, and that of metals with the antimony structure falls by a factor of 2. Table 1 shows that the Perlitz rule is only very roughly obeyed by a number of metals; there

TABLE 1. Ratio of the Electrical Conductivities in the Solid
and Liquid States at the Melting Point

Metal	z_s	z_l	ΔV, %	$(\sigma_s/\sigma_l)_{exp}$	$(\sigma_s/\sigma_l)_{calc}$
Mercury	6+6+6	6+4	3.5	3.4	2.23
Gold	12	11	5.2	2.28	2.22
Zinc	12	11	6.5	2.24	2.3
Aluminum	12	11	6.4	2.2	2
Indium	12	8−4	2.7	2.18	—
Tin	4+2	11	2.7	2.1	3
Silver	12	—	4.5	2.09	2
Thallium	12	8+4	3.2	2.06	2.3
Copper	12	—	4.1	2.04	1.97
Lead	12	8+4	3.4	1.94	1.87
Cadmium	12	8+4	4.7	1.93	2.3
Magnesium	12	—	4.2	1.78	—
Lithium	8	9	1.5	1.68	1.5
Cesium	8	—	2.5	1.65	1.75
Rubidium	8	—	2.5	1.60	1.76
Potasium	8	8	2.5	1.56	1.75
Sodium	8	8	2.5	1.45	1.58
Nickel	12	—	—	1.3	2.34
Cobalt	12	—	—	1.05	—
Iron	12	—	—	0.9	1.67
Antimony	3	—	−0.95	0.61	5.6
Manganese	—	—	—	0.6	—
Gallium	1	11	−3.4	0.58	4.5
Bismuth	3	7−8	−3.3	0.45	5.0

is almost a continum of values of σ_s/σ_l, and some of the later investigations contradict altogether the Perlitz rule. For example, nickel, cobalt, and iron have a close-packed structure and their resistance should double on melting while in fact the values of σ_s/σ_l for these metals are 1.3, 1.05, and 0.9, respectively. Therefore, Perlitz's assumption of a random distribution of atoms in the melt is not related to the small change in the electrical conductivity on melting and is contradicted by the results of later investigations of the structure of liquid metals (Chap. III).

Mott started from the opposite viewpoint [18]. He assumed that the atoms in a liquid have a regular distribution, which although less perfect than

in a solid, is sufficiently ordered to justify the term "liquid crystals." Therefore, the motion of an electron in a liquid metal is very similar to its motion in a solid and the following formula deduced for crystals also applies to
liquid metals:

$$\sigma = 2.83 \cdot 10^{-32} \frac{n M_A}{C^2} \frac{\theta^2}{T} \quad \text{e.s.u.;} \qquad (1.1)$$

here n is the electron density; M_A is the atomic weight; θ is the Debye temperature; T is the temperature;

$$C = \frac{\hbar^2}{2m} \int |\nabla u_{\mathbf{k}}|^2 \, d\mathbf{r} + \int |V| \, u_{\mathbf{k}}|^2 \, d\mathbf{r}; \qquad (1.2)$$

m is the electron mass; $u_{\mathbf{k}}$ is a modifying factor in Bloch's function (Chap. II);
V is the atomic potential; the integrals are taken over a unit cell.

Mott assumed that melting solely alters n and θ in Eq. (1.1); n changes
only by 5%, so that the change in σ is mainly due to the change in θ, which
can be calculated as follows. At the melting point, the partition functions
(sums-over-states) of the solid Z_s and liquid Z_l are equal and, according to
statistical physics, their ratio is

$$1 = \frac{Z_l}{Z_s} = e^{-\frac{\Delta E}{k_B T}} \left(\frac{1 - e^{-\frac{\hbar \nu_s}{k_B T}}}{1 - e^{\frac{\hbar \nu_l}{k_B T}}} \right)^3 ; \qquad (1.3)$$

where ν_s and ν_l are the atomic vibration frequencies in the solid and liquid
states. The actual frequency distribution is ignored; it is assumed that all
atoms vibrate at the same frequency $\nu_s = k_B \theta_s / \hbar$ or $\nu_l = k_B \theta_l / \hbar$. The difference between the rest energies of atoms in a solid and liquid, ΔE, can be
found from the latent heat of fusion L using the following relationship

$$\Delta E = L + \frac{3\hbar \nu_s}{e^{\frac{\hbar \nu_s}{k_B T_{mp}}} - 1} - \frac{3\hbar \nu_l}{e^{\frac{\hbar \nu_l}{k_B T_{mp}}} - 1} ; \qquad (1.4)$$

T_{mp} is the melting point temperature. If $T_{mp} \gg \theta$, then $k_B T_{mp} \gg h\nu$ and we
find approximately from Eqs. (1.3) and (1.4)

$$\frac{\sigma_s}{\sigma_l} = \left(\frac{\theta_s}{\theta_l}\right)^2 = \left(\frac{\nu_s}{\nu_l}\right)^2 = e^{\frac{2L}{3k_B T_{mp}}}. \tag{1.5}$$

The values of σ_s/σ_l calculated using the above formula [19] are listed in the last column of Table 1. Mott's theory accounts approximately for the values of σ_s/σ_l of the majority of metals but does not reproduce the variation of this ratio from one metal to another and is completely incapable of explaining the increase in the electrical conductivity on fusion, which is observed in some metals.

A weakness of Mott's theory is the assumption that the parameters n and C do not change on fusion. C is very sensitive to any change of volume and the mutual positions of the atoms; n changes when the short-range order alters on fusion. The reduction of the resistance of antimony, manganese, bismuth, and gallium on melting is related to such a change of the short-range order. The density of these three metals also increases on melting. Cusack and Enderby [20] ascribed the change in the electrical conductivity of metals on melting to a change in their entropy, which can be divided into vibrational and structural parts. In different metals, one or the other part may dominate.

Busch and Vogt [21] measured the electrical conductivity and the Hall effect as a function of temperature below and above the melting point. They found that the Hall coefficient decreased by a factor of 9 on fusion, indicating that the electron density n rose by a factor of 9. If we take into account this increase of n, a satisfactory agreement is obtained with Mott's formula for the change in σ on melting.

Regel' [22], compared the data on the electrical conductivity and the Hall coefficient of solid and liquid metals and concluded that, on melting, the electron mobility in metals changed very little. Cusack and Kendall[23] concluded, from measurements of the Hall effect in liquid mercury, that the Hall coefficient approached the free-electron-gas value after melting.

The Hall effect in liquid metals has also been recently measured by Busch and Tièche [24].

Table 2 lists data on the temperature dependence of the resistivity of thirteen metals investigated by Roll and Motz [2] and three metals investigated by Mokrovskii and Regel' [1]. The last column of Table 2 gives the temperature coefficient of the resistivity

$$\beta = \frac{1}{\rho_{T_{mp}}} \cdot \frac{d\rho}{dT}. \tag{1.6}$$

TABLE 2. Resistivity of Molten Metals as a Function of Temperature (in 10^{-6} ohm · cm)

Metal	T_{mp}, °C	Temperature, °C										β, 10^{-5}/°C
		300	400	500	600	700	800	900	1000	1100	1200	
Aluminum	658	–	–	–	–	24.7	26.2	27.7	29.2	30.6	32.1	59
Antimony	630	–	–	–	–	115.4	118.1	120.8	123.5	–	–	25
Lead	327	–	98.2	102.9	107.6	112.3	116.9	121.6	126.3	–	–	48
Gold	1063	–	–	–	–	–	–	–	–	31.8	33.1	45
Indium	154	36.7	39.3	41.9	44.4	47.0	49.6	52.2	54.7	–	–	77
Cadmium	321	–	34.7	35.2	36.3	–	–	–	–	–	–	–
Copper	1083	–	–	–	–	–	–	–	–	21.2	22.1	42
Magnesium	650	–	–	–	–	27.7	28.2	28.7	–	–	–	15
Silver	960	–	–	–	–	–	–	–	17.6	18.4	19.3	52
Thallium	302	–	76.2	79.1	81.9	84.8	87.7	–	–	–	–	40
Bismuth	271	131.9	137.6	143.3	149.0	154.7	160.4	166.1	171.8	–	–	44
Zinc	419	–	–	36.8	36.3	36.4	36.3	–	–	–	–	–
Tin	232	49.7	52.5	54.7	57.2	59.6	62.1	64.5	67.6	69.5	72.0	52

Nickel $\rho_t = 85 \,[1+1.5 \cdot 10^{-4}\,(t-1450)]$ $(1450 \leq t \leq 1700)$

Cobalt $\rho_t = 102 \,[1+6 \cdot 10^{-4}\,(t-1490)]$ $(1490 \leq t \leq 1700)$

Iron $\rho_t = 110 \,[1+3 \cdot 10^{-4}\,(t-1535)]$ $(1535 \leq t \leq 1700)$

In the majority of cases, with the exception of cadmium and zinc, the resistivity varies linearly with temperature. If the mechanism of electron scattering in liquid metals is exactly the same as in solid metals, then according to Eq. (1.5), we should have

$$\beta = \frac{1}{T_{mp}}. \qquad (1.7)$$

The experimental values of β are almost always smaller. As pointed out by Roll and Motz [2], only copper, silver, gold, and aluminum have approximately the same $d\rho/dT$ for the solid and liquid states. All other metals have a lower value of $d\rho/dT$ in the liquid than in the solid state. This indicates the presence of nonphonon scattering of electrons in molten metals, which is due to departures from the short-range order. On the other hand, the experimentally measured resistivity rise of some metals is smaller than that predicted by Mott's theory (Table 1). A comparison of these two observations indicates that for these metals Mott's theory predicts too high a value of the phonon contribution to the electrical resistance of liquid metals.

§2. Changes in the Electrical Conductivity of Semiconductors on Melting

Until recently, there have been no data on the electrical properties of liquid semiconductors. Only the electrical conductivity of tellurium [25] and the electrical conductivity and the thermoelectric power of liquid selenium [26, 27] were known.

Extensive and detailed investigations were carried out by Blyum, Mokrovskii, and Regel' [28, 34], who measured the electrical conductivity of a large number of semiconductors as a function of temperature, passing through the melting point. Measurements both in the solid and liquid states were carried out without using electrodes, which was more convenient at higher temperatures — in spite of the reduced accuracy — since the difficulty of the selection of electrodes stable over a wide range of temperatures was avoided. The electrical conductivity was measured in a rotating magnetic field which acted on a sample in the same way as on the armature of an asynchronous motor and caused twisting of the elastic filament by which the sample was suspended. The angle of twist was a measure of the electrical conductivity of a sample of given dimensions.

Individual semiconductors have been investigated in the solid and liquid states by other workers [35-37].

We shall briefly review the results of measurements for various semiconductors.

1. Germanium [29, 35] is a typical semiconductor in the solid state; its lattice is of the diamond type and near the melting point its intrinsic conductivity has a temperature dependence represented by a forbidden band width $E_g = 0.75$ eV. On melting, the electrical conductivity of germanium rises suddenly by a factor of 13 and becomes metallic. The value of this conductivity is typical of a liquid metal; on increase of temperature, the resistance remains constant at first and then begins to rise.

2. Silicon [27] is also a diamond-type semiconductor in the solid state. Above 800°C up to the melting point (1420°C), the conductivity is intrinsic and its rise with temperature is represented by a forbidden gap $E_g = 1.05-1.12$ eV. On melting, the electrical conductivity rises by a factor of 29 and becomes metallic; on further heating, the resistance rises.

3. InSb [29, 35] has the structure of sphalerite; in the solid state, it is an intrinsic semiconductor with $E_g = 0.15$ eV. On melting (540°C), the electrical conductivity increases suddenly by a factor of 2.8 and then decreases considerably on further heating.

4. GaSb [30, 32, 33, 35] is also a semiconductor with the sphalerite structure. In the solid state, its electrical conductivity depends strongly on the sample's purity and weakly on temperature; the conductivity curve has inflections only near the melting point.

On melting, the electrical conductivity of GaSb rises suddenly, the extent of this rise depending on the sample's purity; the maximum change of the electrical conductivity is by a factor of 30. In the liquid state, the electrical conductivity is practically the same for samples of different purities and is independent of temperature.

5. HgSe [29, 30] has the sphalerite structure. The electrical conductivity depends on the concentration of excess selenium. In the solid state, the metallic type of temperature dependence of the resistivity is observed but on melting the resistivity rises discontinuously by a factor of 20. Above the melting point, the temperature dependence of the resistivity is of the semiconductor type, corresponding to $E_g \simeq 2.3$ eV.

6. HgTe [29, 32] has the sphalerite structure. Like HgSe, it has the metallic type of temperature dependence of the resistivity in the solid state,

but on melting the resistivity drops sharply and further heating reduces it even more.

7. Gray selenium [29] has a complex lattice, consisting of chains with covalent bonds along them. Selenium is a semiconductor both in the solid and liquid state, but on melting the resistivity rises by a factor of 3000 and the semiconducting properties of liquid selenium are much more prominent than those of selenium crystals.

8. Solid tellurium [29, 34, 36] also has a complex structure, consisting of chains with covalent bonds, but the behavior of its electrical properties is quite different from that of selenium. In the solid state, tellurium is a semiconductor and its resistance decreases with temperature. On melting, the resistance decreases by a factor of 16 and continues to decrease beyond the melting point but beginning at 700°C the resistance becomes independent of temperature.

9. Solid PbTe [33] exhibits metallic conduction at room temperature, and semiconducting properties at temperatures above 300-500°C. The behavior of the conductivity depends strongly on heat treatment and sample purity. On melting, the electrical conductivity rises discontinuously by a factor of 3-4 but on further heating still retains the semiconducting nature. A similar dependence of the electrical conductivity on temperature is also observed for PbSe and several other substances.

10. Cu_2Se [32, 33] has a complex cubic lattice with two different positions of the copper atoms. The electrical conductivity is practically unaffected by melting, retaining the same small temperature coefficient in the solid and liquid state, behaving throughout as a high-conductivity semiconductor.

11. Cu_2Te [32] has the same structure as Cu_2Se. In the liquid and solid states it is a high-conductivity semiconductor with the same temperature coefficient of the resistivity. The resistivity rises somewhat on melting.

12. Tl_2S [34] is a typical semiconductor both in the liquid and solid states with the same slope of the dependence of the logarithm of the electrical conductivity on temperature. On melting, its electrical conductivity rises somewhat.

13. Bi_2O_3 [34] is also a typical semiconductor both in the solid and liquid states; the temperature dependence of $\ln \sigma$ is almost a straight line, with a small peak near the melting point. Investigations of the thermoelectric power and resistivity have shown that below 500°C n-type conductivity predominates and that its contribution increases with rise of temperature.

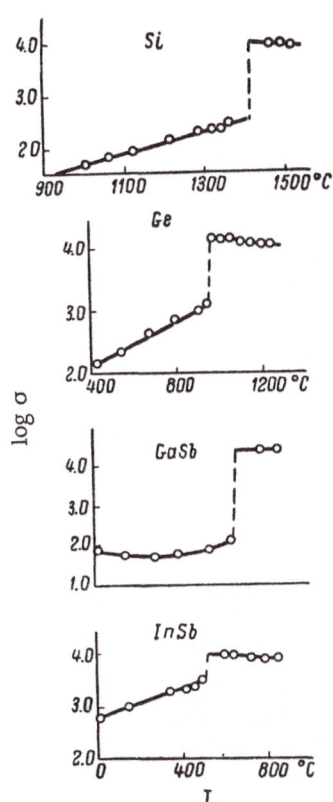

Fig. 1. Temperature dependend-
ence of the electrical conductiv-
ity (σ, in ohm$^{-1} \cdot$ cm^{-1}) of silicon,
germanium, GaSb, and InSb in the
solid and liquid states.

14. Sb_2S_3 [34] has a semiconductor type of temperature dependence of ln σ in both the liquid and
solid states, but in the liquid state the
slope of the straight line is somewhat less. On melting, the electrical
conductivity rises quite sharply by a
factor of several times.

15. V_2O_5 [37] also exhibits a
semiconductor type of electrical conductivity both in the liquid and solid
states, but in contrast to Tl_2S and Sb_2S_3
the electrical conductivity falls sharply
by a factor of several times on melting.

16. V_2O_5–CuO alloys in the liquid state have a conductivity which
rises with temperature, the rise being
stronger for alloys with a bigger pro-
portion of CuO. On melting, the conductivity drops sharply, like that of
pure V_2O_5.

17. FeS–Cu_2S alloys in the liquid state exhibit a gradual change in
the temperature dependence of the
electrical conductivity with increase
of the Cu_2S content. In pure FeS, the
electrical conductivity decreases with
temperature; in alloys with 25-35%
Cu_2S, it is practically independent of
temperature; and when the Cu_2S content is more than 50%, the conductiv-
ity increases with temperature and is represented by a straight line with a kink.

Many oxides and sulfides of metals have the typical semiconducting
temperature dependence of the electrical conductivity with approximately the
same activation energy in the solid and liquid states. Sometimes, an ionic
conduction contribution is observed.

It follows that, in contrast to metals, the behavior of semiconductors on
melting is more differentiated and the magnitude of the electrical conductivity

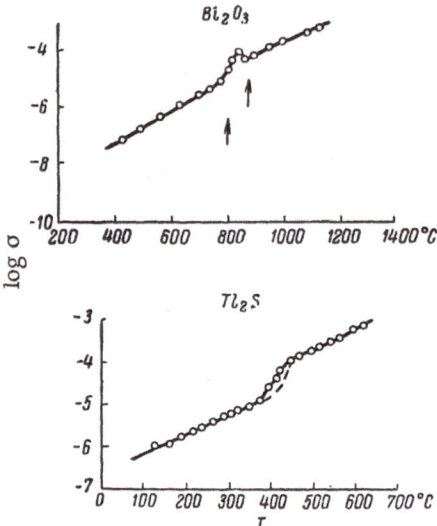

Fig. 2. Temperature dependence of the
electrical conductivity of Bi_2O_3 and Tl_2S
in the solid and liquid states. (The verti-
cal arrows in Figs. 2 and 3 denote the melting points.) In the upper half of Fig. 2, the
left-hand arrow represents yellow Bi_2O_3, and
the right-hand one, white Bi_2O_3.

frequently changes very considerably. Two main groups may be distinguished in semiconductors.

   1. Substances which are semiconductors in the solid state and become
metals after melting. Their electrical conductivity rises very sharply on melting. Among these substances are elements of group IV, compounds of the
$A^{III}B^V$ type, which crystallize in diamond or sphalerite lattices (Ge, Si, InSb),
and tellurium, which has a chain structure. In all these substances, the coordination number increases on melting. For example, x-ray diffraction studies
of germanium [38] have shown that its coordination number increases from 4
to 8 on melting. We may assume that the same occurs in other substances with
the same structure as germanium. Figure 1 shows the temperature dependence
of the electrical conductivity of semiconductors of this group — silicon, germanium, GaSb, and InSb — in the solid and liquid states.

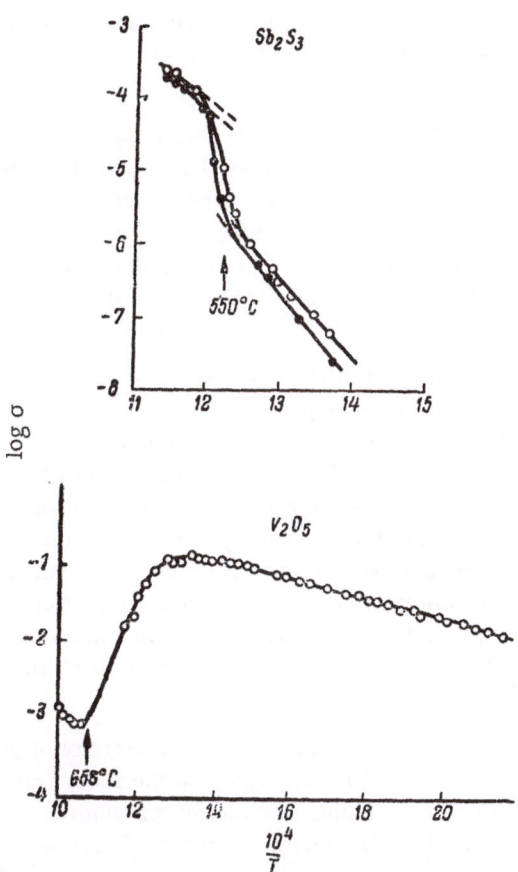

Fig. 3. Temperature dependence of the electrical conductivity of Sb_2S_3 and V_2O_5 in the solid and liquid states.

2. Substances which are semiconductors both in the solid and liquid state. On melting, their electrical conductivity changes but little and the same temperature dependence of the conductivity is retained above the melting point. This group includes practically all metal oxides and sulfides, and the majority of selenides and tellurides, particularly of light metals. In these substances, the coordination number is not altered by melting, i.e., the short-range order is only slightly affected.

Figure 2 shows, by way of example, the temperature dependence of the electrical conductivity of two substances of the second group, Tl_2S and Bi_2O_3, whose conductivity and its temperature dependence change very little on melting. Figure 3 shows the temperature dependence of the electrical conductivity of two other substances, Sb_2S_3 (from two sources) and V_2O_5, which are of practical interest. Their electrical conductivity exhibits a small jump on melting, but the temperature dependence remains the same beyond the melting point.

In each group, there are considerable differences in the behavior of the electrical conductivity on melting, which are due to differences in the change of structure on melting. Several substances cannot be classified in either of these two groups, for example, HgSe, HgTe, PbSe, and PbTe. The metallic type of conduction is exhibited by HgSe and HgTe right up to the melting point, and by PbSe and PbTe at relatively low temperatures. In the liquid state, all these compounds behave as semiconductors. This may be due to the fact that in the molten state they have molecular structures and dissociate on increase of temperature, which gives rise to increased electrical conductivity. The lowest degree of dissociation occurs in HgSe, whose resistivity "jumps" on melting.

The sharp reduction of the electrical conductivity of gray selenium is explained by the loosening of its structure and the reduction of the intensity of interaction between the chains, i.e., liquid selenium is a molecular liquid although it consists of the same chains as in the solid state.

The reviewed experimental data lead to the important conclusion that the properties of semiconductors are governed by their short-range order and by the nature of the bonds between atoms. On melting, the electrical conductivity changes greatly in those cases when the short-range order is radically altered. A change in the long-range order without a change in the short-range order is accompanied by slight changes in the electrical conductivity and its semiconductor nature is retained above the melting point.

§ 3. Correlation Between the Electrical Conductivity and Other Properties

To gain a better understanding of conduction, it is useful to consider the correlation between the conductivity and other properties of liquid conductors, such as the density, viscosity, thermoelectic power, etc.

The relationship between the changes in electrical conductivity and density of melting metals can be seen from Table 1. A greater increase in volume on melting, ΔV, corresponds, on the average, to a higher value of the ratio σ_s/σ_l; for metals with negative ΔV, the ratio σ_s/σ_l is less than unity.

Mokrovskii and Regel' [39] measured the density of several semiconductors which crystallize with diamond or sphalerite lattices (Ge, GaSb, InSb, HgSe, and HgTe), both in the solid and liquid states, with the particular aim of establishing the correlation between the change in density and electrical conductivity. The results of their measurements have shown that such a correlation does indeed exist.

In the case of Ge, GaSb, and InSb, the electrical conductivity rises sharply on melting and the density correspondingly: by 5.35% in Ge, 7.5% in GaSb, and 12.9% in InSb. The electrical conductivity of HgSe falls sharply on melting and, correspondingly, the density falls by 6.3%. Finally, the change in the electrical conductivity of HgTe on melting is complex and there is a complex dependence of the density on temperature.

The magnitudes of the jumps of the electrical conductivity and density are not related in a regular fashion. The greatest increase in the density on melting is observed for InSb but its electrical conductivity rises only by a factor of 2.8; this is related to the relatively high conductivity of InSb in the solid state.

Observations of the change in the density confirm the hypothesis of changes in the structure of these substances. In GaSb and InSb, as in Ge, the coordination number increases considerably on melting. Conversely, in HgSb, a reduction in the density indicates a reduction in the coordination number to 2 or even 1 (molecular liquid).

On a different occasion, Mokrovskii and Regel' [40] measured the density of liquid alloys of tellurium and selenium. The temperature dependences of the specific volume of these alloys exhibit a minimum which lies at 450°C in the case of pure tellurium but shifts to higher temperatures on the addition of selenium. Mokrovskii and Regel' point out that this shift correlates well with the temperature dependence of the electrical conductivity of these alloys. The singularities of the temperature dependence of the specific volume and the electrical conductivity are due to the disruption of the chain structure with increase in temperature.

Blyum and Regel' investigated the thermoelectric power [41] and viscosity [42] of tellurium and selenium alloys in the liquid and solid states. The temperature dependence of the thermoelectric power was investigated for tellurium and two of its selenium alloys (85% Te + 15% Se and 30% Te + 70% Se) in the range from 100 to 500°C. It has been found that, on melting, the thermoelectric power decreases sharply, which is in agreement with the increase in the electrical conductivity and indicates that the carrier density becomes higher and the bonds become more metallic. On melting, the thermoelectric power of tellurium falls to 10 μV/deg (relative to copper) and then remains constant up to 500°C; its sign represents p-type conduction. Alloys of tellurium with selenium have more complicated temperature dependences of their thermoelectric powers, which rise with temperature above the melting points. On increasing the selenium content of such an alloy, its thermoelectric power near the melting point increases both in the solid and liquid states.

The viscosity was measured as a function of temperature for several alloys of various compositions, ranging from pure tellurium to pure selenium. From the results of these measurements, it follows that there is a correlation between the behavior of the resistivity and the viscosity of the samples, both as a function of temperature for a sample of certain composition, and as a function of composition at a fixed temperature. This correlation is disturbed somewhat at higher temperatures for samples containing more than 10% selenium. In the case of tellurium and samples containing small amounts of selenium, the transition through the melting point is accompanied by a sharp drop in the viscosity and resistivity. On increasing the selenium content, the change in the properties becomes smoother and in a sample containing 70% selenium, there is practically no discontinuity in the electrical conductivity. In the case of selenium, as just mentioned, the resistivity rises sharply at the melting point and then gradually decreases with temperature. The viscosity of selenium also decreases continuously over a wide range of temperatures.

Above 500°C, the viscosity depends little on temperature, decreasing slowly as the temperature rises. The resistivity of samples with a low selenium content also decreases with rising temperature in this region.

The results of the viscosity measurements confirm that the chain structure of tellurium is destroyed on melting. Increase in the selenium content makes this more difficult and slows down the reduction in the viscosity and resistivity.

§4. Characteristics of the Electrical Conductivity

of Liquid Solutions of Metals and Semiconductors

A typical dependence of the electrical conductivity on the composition of solid metal alloys is a curve with a broad minimum near the 50% composition. Similar dependences have been observed for some systems in the molten state, for example, sodium—potassium alloys. The linear dependence of the electrical conductivity on composition is interpreted as indicating the absence of a uniform atomic solution and the presence of a mechanical mixture of macroscopic volumes of the components. Later investigations have shown, however, a considerably different behavior of the electrical conductivity of several solutions.

Ablova, Elpat'evskaya, and Regel' [43] investigated germanium and silicon alloys of various compositions from 0 to 48.6 molar % of silicon. The following properties were measured as a function of composition: the forbidden band width E_g; the ratio of the conductivities in the liquid and solid states at the melting point σ_l/σ_s; and the electrical conductivity of liquid solutions σ_l. The results were supplemented by measurements on pure silicon. It was found that the forbidden band width rises monotonically with the silicon concentration, the main rise occurring at low concentrations.

The curves for σ_s/σ_l and σ_s have sharp maxima at the 50% composition, which is the opposite of the behavior of "typical" metallic alloys. Such unusual behavior is, of course, related to a basic change in the short-range order on melting. We may assume that at the 50% composition, the coordination number changes most and the bonds become metallic, giving rise to the maximum conductivity.

Gaibullaev and Regel' [44] measured the electrical conductivity of the continuous atomic solutions Ag—Au, In—Sb, and Bi—Sb, as well as liquid eutectic systems of a large number of metal pairs [45].

In the solid state, these three systems exhibit the usual dependence of the electrical conductivity on composition, but as the temperature increases, the conductivity curve becomes straighter and the minimum disappears. For Ag—Au solutions, $\sigma_s/\sigma_l > 1$ and depends very weakly on composition; for In—Sb, σ_s/σ_l is also greater than unity, but depends strongly on composition; for Bi—Sb, $\sigma_s/\sigma_l > 1$ and the curve representing this ratio has a maximum at the 50% composition, as in the germanium—silicon case, but this maximum is not so sharp. The results obtained show that in Bi—Sb solutions the covalent bonds are destroyed and the metallic ones are strengthened, but these effects are weaker

in Bi—Sb solutions than in Ge—Si solutions and, therefore, the dependence of
the electrical conductivity on composition in Bi—Sb is intermediate between
the dependences observed for Ge—Si and for typical metals.

Gaibullaev and Regel' [45] also measured the electrical conductivity of
the eutectics of the systems Au—Sn, Pb—Te, Ge—Te, Sn—Te, Cu—Sb, Bi—Te,
Pb—Sn, Bi—Cd, Cd—Zn, Bi—Sn, Bi—Pb, Sn—Zn, Cd—Pb, Pb—Sb, Ag—Sb, and
Ag—Sn, from room temperature to 1260°C. Three main types of temperature
dependence of the resistivity are observed.

One type is represented by the Pb—Sn and Bi—Cd systems, which obey
the rule of additivity of the electrical conductivities of the components of a
mixture over the whole temperature range. This rule has been established theoretically for systems which are mechanical mixtures of components of very
low mutual solubility. Thus for these systems the nature of the eutectic mixture is retained in the liquid state up to high temperatures, i.e., there is a ten-
dency for atoms of one type to aggregate in large uniform regions.

Another type is represented by systems such as Cd—Zn, Cd—Sn, Bi—Sn,
and Bi—Pb, for which the temperature dependence of the resistivity of liquid
systems of eutectic composition has no singularities but differs considerably
from the calculated curve. This difference is explained qualitatively by the
limited mutual solubility of the components, which increases,with temperature,
and eventually — hundreds of degrees above the melting point of the eutectic —
gives rise to homogeneous atomic mixtures.

The third type of temperature dependence is represented by the systems
Sn—Zn, Cd—Pb, Pb—Sb, Ag—Sb, and Ag—Sn, which exhibit marked singularities of the temperature dependence of the resistivity. These singularities
represent a rapid change of the structure (short-range order) of the alloy due
to the mutual solubility of the components and the transition to homogeneous
atomic solutions. Thus, studies of the electrical conductivity of all these eutectic systems indicate that the eutectic nature of the structure may be retained in the molten state over a wide or narrow range of temperatures.

§5. Formation of the Glassy State in Semiconducting

Systems

Many substances may exist in the glassy state. The glasses which have
been investigated most intensively are those which are alloys of the oxides:
SiO_2, Na_2O, CuO, PbO, etc. These glasses are excellent insulators with very

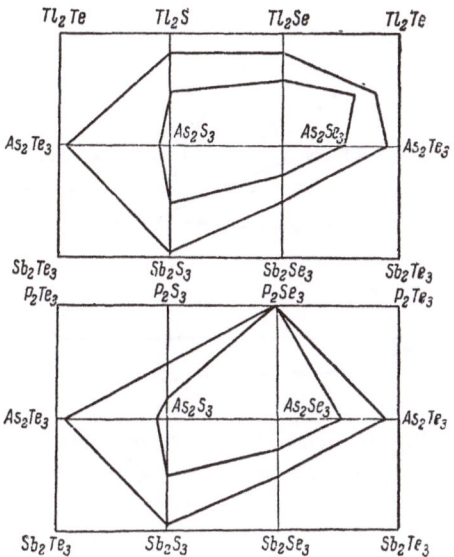

Fig. 4. Glass-formation regions in systems of thallium, arsenic,and antimony chalcogenides (upper half) and phosphorus, arsenic and antimony chalcogenides (lower half). The inner polygon represents slow-cooling conditions, the outer one rapid quenching.

weak ionic conductivity. Information on glassy semiconductors has, until recently, been very scarce. Amorphous selenium, exhibiting semiconducting properties, is known best.

Systematic investigations of glass formation in several semiconducting systems were carried out by Goryunova and Kolomiets [46-48], who studied the change in the electrical properties of $Tl_2Se-Sb_2Se_3$ upon the isomorphous substitution of one element by another [46]. They discovered that when antimony was replaced completely with arsenic, $Tl_2Se-As_2Se_3$ was found to be an amorphous semiconductor. The introduction of tellurium increased the conductivity to $\sigma = 10^{-3}$ ohm^{-1}· cm^{-1}. The isomorphous substitution of selenium with tellurium gave rise to a new group of amorphous substances with compositions varying from $Tl_2Se-As_2Se_3$ to $Tl_2Te-As_2Te_3$.

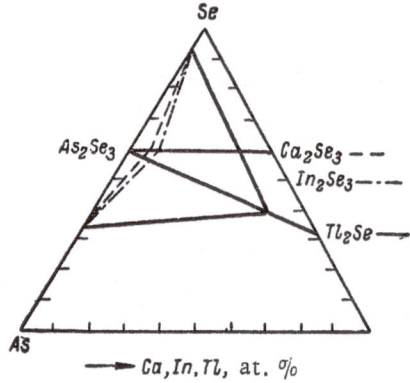

Fig. 5. Glass-formation regions in the
system $As-Se-A^{III}$.

The same workers also investigated the systems [47]: 1) $Tl_2Se[xAs_2Se_3 \cdot (1-x)Sb_2Se_3]$ and 2) $xAs_2Se_3 \cdot (1-x)Sb_2Se_3$.

In both of these systems, the conductivity is of the semiconductor type, although it depends strongly on composition. The initial material for the first system — $Tl_2Se \cdot Sb_2Se_3$ — is crystalline but as the antimony is replaced with arsenic, the crystallinity becomes less and less marked until $Tl_2Se \cdot As_2Se_3$, an amorphous glassy substance, is formed. Rapid cooling gives glassy alloys of the composition $Tl_2Se[(\frac{5}{6})As_2Se_3 \cdot (\frac{1}{6})Sb_2Se_3]$ and $Tl_2Se[(\frac{2}{3})As_2Se_3 \cdot (\frac{1}{3})Sb_2Se_3]$.

These substances can be transformed from the crystalline to the amorphous state, and conversely, by quenching and annealing.

In the second system, the alloys As_2Se_3 and $(\frac{5}{6})As_2Se_3 \cdot (\frac{1}{6})Sb_2Se_3$ are glass, but all the remaining alloys are crystalline. The transition to the glassy state is accompanied by a sharp drop in the electrical conductivity.

Further studies [48] have shown the existence of similar glassy materials in the systems: 1) $As_2Se_3-Tl_2Se_3$; 2) $As_2Se_3-As_2Te_3$; 3) $As_2Se_3-As_2S_3$; and 4) $As_2S_3-Sb_2S_3$.

The establishment of the glassy state was indicated by the following: conchoidal fracture, absence of halts in the cooling curves, and the ability to be drawn into filaments or to spread into thin films. X-ray diffraction investigations confirmed that the amorphous substances obtained are typical glasses.

TABLE 3. Viscosity of Glasses of the As_2Se_3–As_2Te_3 System

As_2Se_3		$4As_2Se_3 \cdot As_2Te_3$		$3As_2Se_3 \cdot As_2Te_3$		$2As_2Se_3 \cdot As_2Te_3$		$As_2Se_3 \cdot As_2Te_3$	
t, °C	ν	t, °C	ν	t, °C	ν	t, °C	ν	t, °C	ν
412	15.5	372	20.6	349	19.8	377	6.17	382	3.03
462	3.3	417	5.45	412	3.88	420	2.04	418	1.23
480	2.74	439	3.46	460	1.37	452	0.915	433	0.81
570	0.5	561	0.25	542	0.29	531	0.2	523	0.139
608	0.29	634	0.076	615	0.0815	610	0.06	580	0.055
640	0.18	782	0.0124	827	$6.35 \cdot 10^{-3}$	777	$6.8 \cdot 10^{-3}$	747	$6.6 \cdot 10^{-3}$
690	0.09	—	—	—	—	—	—	—	—

The components of the investigated alloys were all of one chemical type and are known as chalcogenides of the respective elements (thallium, antimony, arsenic, etc.).

Goryunova, Kolomiets, and Shilo [49, 50] determined the regions of glass formation in alloys of phosphorus, arsenic, antimony, bismuth, and thallium chalcogenides for slow and fast cooling from the liquid state, and in complex chalcogenides based on arsenic sulfide and selenide.

In earlier work [49], the objects of investigation were alloys of ten pseudobinary sections of ternary systems: 1) As_2Te_3–As_2S_3; 2) As_2S_3–As_2Se_3; 3) As_2Se_3–As_2Te_3; 4) As_2S_3–Sb_2S_3; 5) As_2Se_3–Sb_2Se_3; 6) Tl_2S–As_2S_3; 7) Tl_2Se–As_2Se_3; 8) As_2S_3–P_2S_3; 9) As_2Se_3–P_2Se_3; 10) P_2Se_3–P_2Te_3; as well as alloys of a pseudobinary section of the system $Tl_2Se \cdot As_2Se$–$Tl_2Te \cdot As_2Te_3$. The results of the investigation are shown in Fig. 4. The inner polygons enclose regions of glass formation for slow cooling at the rate of 1 deg/min, and the outer polygons represent glass formation on cooling at the rate of 200-300 deg/sec. Figure 4 shows that glass formation depends particularly strongly on the cooling conditions in the case of the As_2S_3–As_2Te_3 system.

Glasses were also obtained for alloys of bismuth chalcogenides with arsenic chalcogenides and in alloys of As_2S_3 with chalcogenides of mercury, lead, tin, copper, silver, gallium, indium, and other elements. All these alloys have small regions of glass formation.

In later work [50], alloys of arsenic sulfide and selenide with sulfides and selenides of elements of groups I, II, III, IVb (with the exception of boron, aluminum, carbon, and silicon) were prepared. The alloy composition was selected to fall on the section lines of the ternary systems $Me-X-As$, where Me is an element of groups I–IV, and X is sulfur or selenium. For elements of group I, the systems were represented by $Me_2X-As_2X_3$; for group II, they were $MeX-As_2X_3$; for group III, they were $Me_2X_3-As_2X_3$, and $Me_2X-As_2X_3$; and for group IV, they were $MeX_2-As_2X_3$, and $MeX-As_2X_3$. The limit of glass formation was taken to be the alloy which became glassy when cooled at 200 deg/sec, and crystalline when cooled at 1 deg/min. Figure 5 shows, by way of example, the approximate regions of glass formation in ternary systems $As-Se-A^{III}$. These regions are bounded by inner triangles the bases of which lie on the As–Se side of the large triangle. Similar diagrams were obtained for elements of groups I, II, and IV but the glass-formation regions were considerably smaller, with the exception of germanium alloys. Sulfides gave results similar to those for selenides.

Goryunova and Kolomiets [51] discussed the relationships governing glass formation and concluded from an analysis of the experimental data that glass formation was related to the chemical nature of atoms, the nature of the electron interaction between them, and the short-range order in the molten state. The necessary condition for glass formation is the presence of covalent bonds both in the solid state and in the melt. A metallic nature of the bonds impedes glass formation and therefore semiconductors of group IV and of the $A^{III}B^V$ type exhibit metallic properties in the molten state but do not form glasses. However, the tetrahedral covalent bonds remaining in the melt also impede glass formation.

In order to determine the structure of glassy semiconductors of the $As_2Se_3-As_2Te_3$ system, Kolomiets and Pozdnev [52] measured their viscosity. The measurements were carried out by the method of torsional vibrations, which gave the kinematic viscosity ν in the temperature range 400-800°C (the softening temperatures of these glasses were between 150 and 190°C). The results of the measurements are listed in Table 3, which indicates that the kinematic viscosity varies smoothly with temperature and with composition, decreasing with increase in temperature and with increase in the As_2Te_3 content.

The plots of log ν as a function of temperature and composition are nearly straight lines. It follows that the activation energy of viscous flow is approximately constant.

It is known that glasses with a three-dimensional structure have the highest viscosity, glasses with a layered or chain structure a lower viscosity, and glasses consisting of a disconnected network of sites have the lowest viscosity. The activation energy of viscous flow also decreases when the association in a glass decreases.

The results of measurements of the viscosity of semiconducting glasses of the $As_2Se_3 - As_2Te_3$ system indicate the absence of complicated complexes. Data on the absolute value of the viscosity suggest a chain structure with covalent bonds within the chains and van der Waals forces between the chains, which accounts for the very low softening temperatures of these glasses. Obviously, the As—Te bond is weaker than the As—Se bond; therefore, an increase in the tellurium content shortens the chains and reduces the glass viscosity.

§6. Electrical Properties of Amorphous Films and Glassy Semiconductors

The presence of electronic conduction in some amorphous substances has been proved in several investigations. The first to be investigated was amorphous selenium. Nasledov and Malyshev [53] measured the electrical conductivity of amorphous selenium and its dependence on the electric-field intensity. In weak fields at room temperature, the conductivity was $\sigma = 10^{-10}$ $ohm^{-1} \cdot cm^{-1}$.

Shidlovskii [54] continued these measurements and obtained the straight lines typical of semiconductors for the dependence of the logarithm on the reciprocal of temperature. Spear [55] measured the transit times of electrons and holes in thin amorphous selenium films and deduced the electron mobility $u_1 = (4.7-5.5) \cdot 10^{-3} cm^2 \cdot sec^{-1} \cdot V^{-1}$ and the hole mobility $u_2 = 0.15 \, cm^2 \cdot sec^{-1} \cdot V^{-1}$.

Several semiconductors (germanium, silicon, silicon oxide, SnS, etc.), deposited in the form of thin films [56], may have an amorphous structure and retain semiconducting properties. Taft and Apker [57] have also shown that amorphous films of antimony less than 300 A thick behave as semiconductors with the Fermi level 0.1 V above the filled band.

Yin Shih-tuan and Regel' [58] investigated the temperature dependence of the resistance of amorphous tellurium films and measured their thermoelectric power. From these measurements, they deduced the hole density (10^{17} cm^{-3}) and mobility ($10^{-2} cm^2 \cdot sec^{-1} \cdot V^{-1}$). Both these quantities vary little with temperature. The hole mobility in an amorphous film is considerably

TABLE 4. Electrical Conductivity of Chalcogenide Alloys in Glassy
and Crystalline States

Chemical composition	Conductivity, $ohm^{-1} \cdot cm^{-1}$	
	Glass	Crystal
$2As_2Se_3 \cdot 3As_2Te_3$	$2.5 \cdot 10^{-8}$	100
$As_2Se_3 \cdot 2Tl_2Se$	$7.5 \cdot 10^{-8}$	$7.2 \cdot 10^{-3}$
$2As_2Se_3 \cdot Sb_2Se_3$	$1.0 \cdot 10^{-10}$	$2.5 \cdot 10^{-6}$
$Tl_2Te \cdot As_2Te_3$	$5.3 \cdot 10^{-3}$	0.46
$Tl_2Se[(\frac{5}{6})As(\frac{1}{6})Sb]Se_3$	$2.2 \cdot 10^{-7}$	$2 \cdot 10^{-8}$

lower than in a crystalline film. It has been found also that oxygen and io-
dine impurities markedly accelerate the crystallization process.

Goryunova and Kolomiets [48] investigated the electrical properties of
five glassy semiconducting systems, based on As_2Se_3: 1) $As_2Se_3-As_2Te_3$; 2)
$As_2Se_3-Tl_2Se$; 3) $As_2Se_3-Sb_2Se_3$; 4) $Tl_2Se \cdot As_2Se_3-Tl_2Se \cdot As_2Te_3$; 5) Tl_2Se
$\cdot As_2Se_3-Tl_2Se \cdot Sb_2Se_3$.

The results of their measurements have shown that the electrical conductivity of the listed glasses varies with composition within very wide limits:
from 10^{-13} to 10^{-3} $ohm^{-1} \cdot cm^{-1}$. Glass having the composition $Tl_2Se \cdot As_2Te_3$
has the highest conductivity ($\sigma = 10^{-3}$ $ohm^{-1} \cdot cm^{-1}$), which is 10^8 times higher
than the maximum conductivity of oxide glasses. All the indications are that
the glasses investigated are typical semiconductors. Thus, the temperature

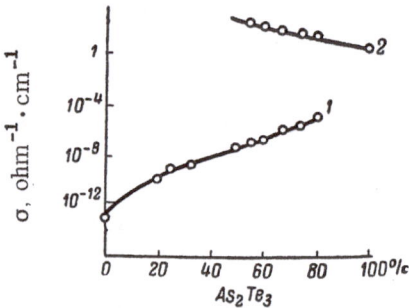

Fig. 6. Dependence of the electrical conductivity on the composition of substances
in the $As_2Se_3-As_2Te_3$ system. 1) Glassy materials; 2) crystalline materials.

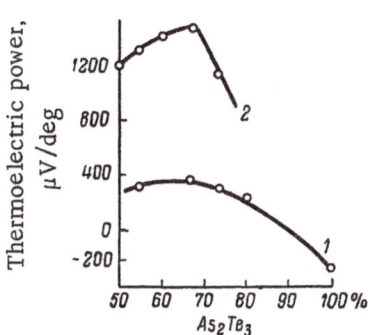

Fig. 7. Magnitude and sign of the thermoelectric power for substances in the $As_2Se_3 - As_2Te_3$ system. 1) Crystalline materials; 2) glassy materials.

dependence of their conductivity obeys the exponential law, all of them exhibit considerable photoconductivity, their conductivity can be n- or p-type, and they exhibit thermoelectric power.

Table 4 lists, for the same composition, the conductivities of some substances in the glassy and crystalline states. In the majority of cases, the transition to the glassy state is accompanied by a considerable drop in the electrical conductivity, but in one case the conductivity rises by almost one order of magnitude.

The properties of materials of the $As_2Se_3 - As_2Te_3$ system were investigated in detail by Vengel' and Kolomiets [59]. They have found that the softening temperatures of glassy materials of this system decrease with increase in the tellurium content and are considerably lower than the melting points of crystalline materials, which increase sharply when the tellurium content is increased.

The density of the alloys also increases sharply with the tellurium content and on the transition from the glassy to the crystalline state (for the same composition).

Figure 6 shows the dependence of the conductivity on composition for glassy and crystalline substances. It indicates that the conductivity of a glass increases by eight orders of magnitude when the As_2Te_3 content is increased. Crystalline materials in the $As_2Se_3 - As_2Te_3$ system have a considerably higher conductivity with a different dependence on composition. The transition of a substance of the same composition from the glassy to the crystalline state is accompanied by a considerable increase in conductivity: for example, the conductivity of $9As_2Se_3 \cdot 11As_2Te_3$ increases by a factor of 10^{10}.

In all samples, both glassy and crystalline, the usual exponential temperature dependence of the conductivity is observed, indicating a semiconducting nature. However, in the case of crystalline samples, the dependence of $\ln \sigma$ on $1/T$ is given by a straight line with a kink, which is typical of impurity semiconductors, while for glass we have a straight line without a change in slope. Special studies have shown that the nature and temperature depend-

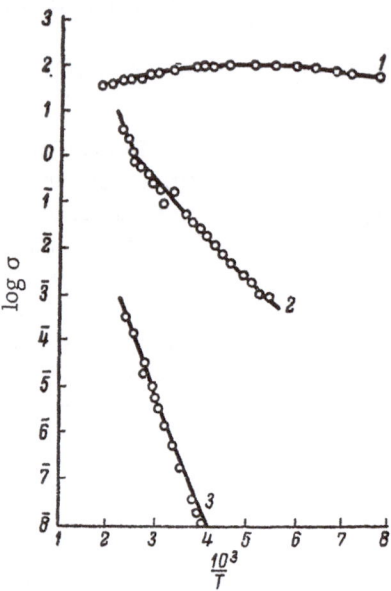

Fig. 8. Temperature dependence of the
electrical conductivity of As_2SeTe_2. 1)
Crystalline material before zone-melting; 2) crystalline material after zone-
melting; 3) glassy material.

ence of the conductivity remain unaltered when glasses melt; the straight line
passes through the softening point without a change in slope.

Figure 7 gives the magnitude and sign of the thermoelectric power as a
function of composition. It is evident that, for the same composition,
glassy substances have considerably higher thermoelectric power than crystalline substances.

A comparison of the temperature dependence of the conductivity of
glassy and crystalline substances indicates that impurities active in crystals
are inactive in glasses. To establish the role of impurities in glassy and crystalline substances, Kolomiets and Nazarova [60] investigated $As_2Se_3 \cdot 2As_2Te_3$,
whose melt crystallized on slow cooling but formed a glass on quenching. The
electrical conductivity of crystalline and glassy samples was measured im-
mediately after preparation and after refining by zone-melting, using the

purification and zone-leveling methods. Zone-melting reduced the conductivity of the crystals by three orders of magnitude (from 60 to 0.04-0.1 $ohm^{-1} \cdot cm^{-1}$) but the electrical properties of glassy As_2SeTe_2 were unaffected by zone melting.

Figure 8 shows the temperature dependence of the electrical conductivity of As_2SeTe_2. The conductivity of crystals before zone-melting varies weakly with temperature but the variation becomes more pronounced after refining and then it represents an activation energy of 0.4 eV. The temperature dependence of the conductivity of the glass is not affected by refining and is represented by an activation energy of 1 eV. It is interesting to note that the activation energies of carriers in glasses and crystals are the same in the intrinsic conduction region. All these observations indicate the absence of impurity conduction in glasses.

Grechanik, Petrovykh, and Karpechenko [61] prepared and investigated glassy oxide semiconductors based on V_2O_5, which were of the $V_2O_5-P_2O_5-RO_x$ type, where RO_x were oxides of various metals. They have established that the electrical conductivity of vanadium glass depends strictly exponentially on temperature in the range from 20 to 250°C, and that the activation energy varies from 0.65 to 1.12 eV depending on composition. No kinks were found in the straight lines log $\sigma = f(1/T)$. We can therefore conclude that these glasses again exhibit intrinsic conduction of the electron — hole type and that impurities are not important. The electrical conductivity of these glasses at 20°C varied from $10^{-4.5}$ to 10^{-12} $ohm^{-1} \cdot cm^{-1}$. The thermoelectric power was also observed.

Ioffe, Patrina, and Poberosvkaya [62] measured the electrical conductivity and thermoelectric power of glasses of the systems $V_2O_5-P_2O_5$ and $V_2O_5-P_2O_5-BaO$. Here again, a rectilinear dependence of ln σ on 1/T was obtained. With increase in the V_2O_5 content, the electrical conductivity increased and the activation energy fell in both systems; this was particularly marked when the ratio of the components was $V_2O_5 : P_2O_5 \leq 2$. The sign of the thermoelectric power suggested n-type conduction. In the temperature range from 300 to 500°K, the thermoelectric power did not depend strongly on temperature. Hence, the authors concluded that the number of carriers (10^{18} cm^{-3}) did not vary with temperature and that the mobility (10^{-2}-10^{-3} cm^2/sec at room temperature) rose exponentially with increasing temperature.

Kornfel'd and Sochava [63] measured the conductivity fluctuations responsible for the "current noise" in solid and liquid antimony sulfide, and in the amorphous semiconductor $Tl_2Te \cdot As_2Te_3$. In the liquid and amorphous substances, the fluctuation level and current noise were very low; for example, in the amorphous modification of $Tl_2Te \cdot As_2Te_3$, the current noise was considerably lower than in the crystalline form. This is due to the fact that in polycrystalline materials the current noise is mainly due to grain boundaries. Thus, amorphous semiconductors are clearly to be preferred in those cases when semiconducting devices must have a low noise level. This advantage will be particularly marked if the device works at low temperatures when the thermal noise level is low.

§7. Photoelectric and Optical Effects in Glassy Semiconductors

Until recently, almost all photoelectric investigations on amorphous semiconductors were confined to selenium.

The most reliable studies were carried out by Gilleo [64], who measured the absorption of light and the photoconductivity of amorphous and hexagonal selenium as a function of temperature. The absorption band edge of amorphous selenium was found to shift from 6600 to 6100 A on cooling from 300 to 90°K. The photoconductivity at low temperatures was observed only for photon energies greater than 2.5 eV. When the temperature was increased the photosensitivity threshold shifted toward longer wavelengths in the same way as the absorption band edge.

The transformation of amorphous into hexagonal selenium led to an increase in the absorption of light in the region from 5000 to 7200 A and to the appearance of a hump at 5200 A at the absorption edge. On cooling the selenium the hump became sharper. The photoconductivity of hexagonal selenium was observed up to 7500 A and its maximum lay at 6300-6400 A, where the absorption was relatively weak, while the photoconductivity maximum of amorphous selenium was at 5600 A, where almost all the light was absorbed.

Somewhat different results were obtained by Billig [65], who reported that the absorption edge of amorphous and crystalline selenium occurred at approximately the same wavelength: 6120 A.

The absorption of light and the photoconductivity of glassy semiconducting chalcogenides were investigated by Kolomiets, Mamontova, and Pavlov [59, 66, 67], mainly for the $As_2Se_3 - As_2Te_3$ system. Their measurements showed that on increase in the As_2Te_3 content the absorption edge gradually shifted toward long wavelengths. All substances of this system exhibited the internal

TABLE 5. Carrier Activation Energies of Glassy Semiconductors

Composition	Activation energy (in eV), determined from		
	absorption band edge	photocon-ductivity	temperature dependence of conduc-tivity
As_2Se_3	1.57	1.64	1.96
$4As_2Se_3 \cdot As_2Te_3$	1.1	1.3	1.46
$3As_2Se_3 \cdot As_2Te_3$	1.05	1.20	1.37
$2As_2Se_3 \cdot As_2Te_3$	1.01	1.04	1.30
$As_2Se_3 \cdot As_2Te_3$	0.94	0.9	1.17
$9As_2Se_3 \cdot 11As_2Te_3$	0.92	0.93	1.17
$2As_2Se_3 \cdot 3As_2Te_3$	0.9	0.83	1.04
$As_2Se_3 \cdot 2As_2Te_3$	0.89	0.83	0.95
$As_2Se_3 \cdot 3As_2Te_3$	0.86	0.74	0.94
$As_2Se_3 \cdot 4As_2Te_3$	0.85	—	0.91

photoeffect and the increase in the As_2Te_3 content broadened the spectral range of the sensitivity and shifted its maximum again toward long wavelengths — from 0.6 to 1.4 μ.

Table 5 lists the carrier activation energies, calculated from the absorption edge, the photoconductivity, and the temperature dependence of the dark conductivity. Table 5 shows the satisfactory agreement between the activation energy values determined from the absorption and the photoconductivity, which were somewhat different from the values found from the temperature dependence of the conductivity. The activation energies calculated in these three ways varied in the same way with composition, i.e., within the experimental error, they decreased monotonically with increase in the As_2Te_3 content.

The variation with composition of glassy substances was the same as for their crystalline modifications.

Measurements carried out on $4As_2Se_3 \cdot As_2Te_3$ at various temperatures[66] showed that the temperature dependence of the photocurrent, and of the dark current, continued smoothly above the softening temperature when the substance changed from the glassy to the liquid state. It was also found that the internal photoeffect also occurred in the liquid state, once again confirming the electronic nature of conduction in chalcogenide glasses.

Kolomiets and Lyubin [68] investigated the photoelectric properties of thin amorphous films of the chalcogenides Sb_2S_3 and As_2Se_3 at various temperatures. They found that the current I varied with the illumination E in accordance with the law $I = AE^n$. At low temperatures and low illumination, $n = 0.9$-1.0; but at high illumination, $n = 0.5$-0.6. Increase in temperature led to an increase in the high-slope region of the lux — ampere characteristic. Maxima were observed in the dependence of the logarithm of the photocurrent on the reciprocal of temperature. These results indicated a complex mechanism of the internal photoeffect in amorphous semiconductors.

§8. Metal — Ammonia Solutions

Another example of liquid electronic conductors are metal—ammonia solutions, i.e., solutions of alkali and alkaline-earth metals in liquid ammonia. Metal—ammonia solutions have for a long time attracted attention because, by varying the concentration of metal atoms, we can vary continuously the electrical conductivity from values representing insulators to values typical of metals.

Experimental work on metal—ammonia solutions is of fairly early origin; a brief review of the results has been given by Vogt [69].

Figure 9 shows a typical dependence of the equivalent electrical conductance of a dissolved metal (sodium) on the degree of dilution of the solution. Beginning from high metallic-type values, the equivalent electrical conductivity rapidly decreases with increase in the average distance between metal atoms, passes through a minimum, and then tends to a limit which represents an infinitely dilute solution. The same figure shows, by a dashed curve, the temperature coefficient of the resistance $(1/R)$ (dR/dT), which has a sharp maximum in the region of the rapid drop in electrical conductivity with the dilution of the solution. Experimental investigations have shown also that the negative charges are the main carriers in metal—ammonia solutions and the mobility of these charges is an order of magnitude higher than the mobility of the positive charges.

At high concentrations of metal atoms, the wave functions of their electrons may overlap and the solution may exhibit metallic conductivity. At lower concentrations of the metal, the solution is a semiconductor with a conductivity due to the metal impurity, and at still lower concentration only polarons may act as carriers.

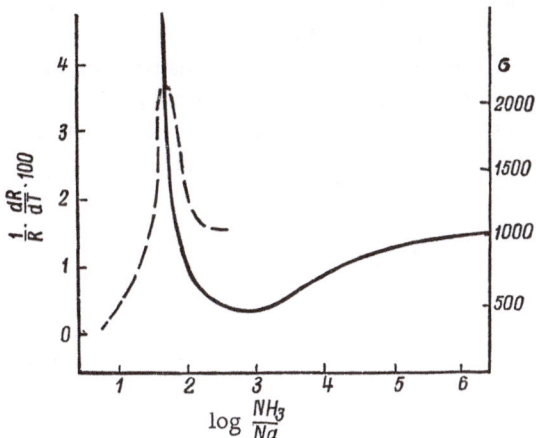

Fig. 9. Dependence of the equivalent electrical
conductance of an ammonia solution of sodium
on the degree of dilution of the solution. The
dashed curve shows the temperature coefficient
of the resistance (1/R) (dR/dT).

The absorption spectra of metal—ammonia solutions exhibit maxima in
the infrared region near 1.8 μ. Red light is also absorbed quite strongly and
this is responsible for the bright blue color of the solutions.

At high dilutions, when the concentration of metal atoms is $n_A < 10^{16}$
cm^{-3}, the solutions are paramagnetic. At concentrations of 10^{17}-10^{18} cm^{-3} the
magnetic susceptibility drops sharply and at $n_A = 10^{20}$-10^{21} atoms/cm^3 and—75°C
the magnetic susceptibility even changes its sign, i.e., the solution becomes
diamagnetic. However, at very high concentrations it is again paramagnetic.

A theory of metal—ammonia solutions has been developed by Deigen
[71]. Using the concepts of local electron centers in an ionic dielectric (polarons, color centers, and double color centers), Deigen explained the special features of the absorption spectra, the dependence of the electrical conductivity
on the concentration of metal atoms, and calculated the positions of the maxima
as well as the half-widths of the bands representing absorption by polarons.
He also extended the theory of the magnetic properties of metal—ammonia solutions, determined the equilibrium concentration and magnetic susceptibility
of local centers, and calculated the susceptibility of the solution as a function
of temperature and metal atom concentration.

Deigen's theory cannot be regarded as a theory of amorphous conductors because it applies equally well to liquids and crystals and it ignores the absence or presence of long-range order in the distribution of atoms. The results of his theory are in agreement with experimental data. This once again shows that there is no basic difference between liquid and crystalline electronic conductors. The mechanism of electron motion in liquids and crystals is of the same nature.

Chapter II

FUNDAMENTALS OF THE ELECTRON THEORY
OF SOLIDS

§9. One-Electron Approximation

In Chap. I we analyzed the experimental data and came to the conclusion that the basic nature of the electron properties of amorphous substances, liquids, and glasses, is the same as that of crystals. Therefore, to construct an electron theory of liquids we shall start with the well-developed and experimentally verified electron theory of solids. The present chapter will review the principal assumptions and methods of this theory.

A solid, like a liquid, is a complex system consisting of many ions and electrons and, strictly speaking, one would solve Schrödinger's equation for this complex system. This cannot be done exactly and many simplifications must be made. In the first approximation, it is usual to neglect the motion of nuclei, on the assumption that their mass is considerably greater than the electron mass. This is an adiabatic approximation, in which the coordinates of nuclei enter Schrödinger's equation as parameters. Next, all electrons are divided into two groups: 1) inner electrons, which belong to filled shells and are tightly bound to nuclei; 2) outer electrons, which are responsible for the majority of the properties of solid substances.

Schrödinger's equation is written down only for the outer valence electrons; the inner electrons, like the nuclei, create only a potential field for the outer electrons.

On these assumptions, Schrödinger's equation contains the following Hamiltonian for n valence electrons:

$$\hat{H} = \sum_{i=1}^{n} \left(-\frac{\hbar^2}{2m} \nabla_i^2 + V_i \right) + \frac{1}{2} \sum_{i,j=1}^{n}{}' \frac{e^2}{r_{ij}} + U_0, \qquad (9.1)$$

where ∇_i applies to the coordinates of the i-th electron; V_i is the potential energy of the i-th electron in the field of nuclei and inner electrons; r_{ij} is the distance between the i-th and j-th electrons; the summation symbol with a prime denotes (here and subsequently) that the terms with i = j are excluded; U_0 is a constant representing the mutual interaction of atomic cores and of the electrons within these cores. This last term is important in determining the binding energy of a solid. In spite of all these simplifications, the exact solution of Schrödinger's equation with the Hamiltonian given by Eq. (9.1) still presents difficulties. The most useful of the approximate methods of solutions is the one-electron approximation, in which the wave function of n electrons is formed from n one-electron wave functions.

In the earliest work it was assumed that the total function of a system may be represented by the product of the functions for individual electrons:

$$\psi(\mathbf{r}_1, \mathbf{r}_2, \ldots, \mathbf{r}_n) = \prod_{i=1}^{n} \psi_i(\mathbf{r}_i); \qquad (9.2)$$

here, $\mathbf{r}_i$ is the radius vector of the i-th electron, and ψ_i is a normalized one-electron function. Hartree [1] suggested that each one-electron function in Eq. (9.2) should satisfy the one-electron Schrödinger equation where the potential includes the Coulomb field of the other $(n-1)$ electrons. This field is calculated by determining the electrostatic potential of an average distribution of electrons, i.e., a field of space charge $|\psi_j|^2$, and is known as the self-consistent field because its value depends on the solution of the following equation:

$$-\frac{\hbar^2}{2m} \nabla_i^2 \psi_i + \left[V_i(\mathbf{r}_i) + \sum_j{}' e^2 \int \frac{|\psi_j|^2}{r_{ij}} \, d\mathbf{r}_j \right] \psi_i = E_i \psi_i, \qquad (9.3)$$

where V_i is the field of the cores, and integration with respect to $d\mathbf{r}_j$ represents integration over the volume associated with the j-th electron. Such an equation is written down for each electron so that we obtain a system of n one-electron equations. Since the Hartree equation includes only the average distribution of the remaining electrons, the correlation in the motion of electrons due to their mutual repulsion is ignored completely. Moreover, the wave function of Eq. (9.2) does not satisfy the Pauli principle according to which the total wave function should be antisymmetric with respect to the transposition of the electron coordinates, including spin variables. Strictly speaking, the functions (9.2) and (9.3) are applicable only to a system of two electrons with antiparallel spins.

In order to statisfy the Pauli principle, we must replace ψ_i with the functions φ_i, which are the products of ψ_i and the spin functions $\eta_i(s_i)$,

$$\varphi(\zeta_i) = \psi_i(\mathbf{r})\,\eta_i(s_i); \tag{9.4}$$

ζ_i are sets of three spatial coordinates and a spin coordinate x_i, y_i, z_i, and s_i. From φ_i, we can construct an antisymmetrical function in the form of the determinant

$$\Psi = \begin{vmatrix} \varphi_1(\zeta_1), & \varphi_1(\zeta_2), & \ldots, & \varphi_1(\zeta_n) \\ \varphi_2(\zeta_1), & \varphi_2(\zeta_2), & \ldots, & \varphi_2(\zeta_n) \\ \cdot & \cdot & \cdot \cdot \cdot \cdot \cdot & \cdot \\ \varphi_n(\zeta_1), & \varphi_n(\zeta_2), & \ldots, & \varphi_n(\zeta_n) \end{vmatrix} = \sum_P (-)^p \, \hat{P} \prod_{i=1}^{n} \varphi_i(\zeta_i), \tag{9.5}$$

in which the elements are $\varphi_i(\zeta_i)$. Here, $\hat{P}$ is the operator representing n! transpositions of n elements, and p is the multiplicity (order) of the transposition, i.e., the number of transpositions necessary to obtain a given result from the initial state. The properties of the determinant ensure the antisymmetric nature of the function Ψ and that the Pauli principle is satisfied because the determinant becomes zero for two equal φ_i.

However, the introduction of antisymmetric wave functions makes Hartree's equation invalid. Therefore, Fock [2] suggested seeking a system of equations of the best one-electron functions using the variation principle. If we start with a function of the type given in Eq. (9.5) and make the condition that the functions φ_i are orthonormalized, i.e.,

$$\int \varphi_i^* \varphi_j \, d\zeta = \delta_{ij}, \tag{9.6}$$

where integration with respect to $d\zeta$ also includes summation over spins, $\delta_{ij} = 0$ if $i \neq j$, and $\delta_{ii} = 1$, Fock's equations for the best functions are:

$$-\frac{\hbar^2}{2m} \nabla^2 \varphi_i + \left[V(\mathbf{r}_1) + \sum_j{}' e^2 \int \frac{|\varphi_j(\zeta_2)|^2}{r_{12}} \, d\zeta_2 \right] \varphi_i(\zeta_1) - $$

$$-\sum_j{}' \left[e^2 \int \frac{\varphi_i(\zeta_2)\,\varphi_j^*(\zeta_2)}{r_{12}} \, d\zeta_2 \right] \varphi_j(\zeta_1) = E_i \varphi_i(\zeta_1) + \sum_j{}' \lambda_{ij}\varphi_j(\zeta_1). \tag{9.7}$$

The derivation of the above equation may be found in [3].

Equation (9.7) differs from Hartree's equation (9.3) by two additional terms. The term in the right-hand part of $\sum_j \lambda_{ij}\varphi_j$, where λ_{ij} are Lagrange's parameters, appears because of the orthogonality condition (9.6) and is not of basic importance. The more important terms are

$$-\sum_j{}'\left[e^2\int\frac{\varphi_i(\zeta_2)\,\varphi_j^*(\zeta_2)}{r_{12}}\,d\zeta_2\right]\varphi_j(\zeta_1),\qquad(9.8)$$

which are called the exchange terms and represent an additional potential which acts on the electrons. This potential does not appear because of some new nonclassical forces, but simply because the total wave function is taken in the form of the determinant (9.5), and not in the form of a simple product (9.3) of one-electron functions. Thus, the exchange terms allow for the correlation in the motion of the electrons, which is due not to the electrostatic repulsion between the electrons but to the Pauli principle. Such a correlation exists only between the electrons with parallel spins.

Adding to the first sum in the left-hand part of Eq. (9.7), the expression

$$e^2\int\frac{|\varphi_i(\zeta_2)|^2}{r_{12.}}\,d\zeta_2\cdot\varphi_i(\zeta_1),$$

subtracting it from the second sum, and assuming that $E_i = \lambda_{ii}$, we can reduce Eq. (9.7) to the form:

$$\hat{H}^F\varphi_i(\zeta_1)=\sum_j\lambda_i\varphi_i(\zeta_1),\qquad(9.9)$$

where $\hat{H}^F$ is Fock's Hamiltonian,

$$\hat{H}^F=-\frac{\hbar^2}{2m}\nabla^2+V+U+\hat{A},\qquad(9.10)$$

U is the Coulomb potential of the distributed charge,

$$U=e^2\int\frac{\rho(\zeta_2,\,\zeta_2)}{r_{12}}\,d\zeta_2,\qquad(9.11)$$

$\rho(\zeta_1,\,\zeta_2)$ is Dirac's density matrix

$$\rho(\zeta_1,\,\zeta_2)=\sum_j\varphi_j^*(\zeta_1)\,\varphi_j(\zeta_2).\qquad(9.12)$$

The operator $\hat{A}$, acting on the function $\varphi_i(\zeta_1)$, represents multiplication of this function by $\rho(\zeta_2, \zeta_1)$, transposition of the variables ζ_1 and ζ_2, and integration with respect to ζ_2. In other words,

$$\hat{A}\varphi_i(\zeta_1) = -e^2 \int \frac{\rho(\zeta_2, \zeta_1) \varphi_i(\zeta_2)}{r_{12}} d\zeta_1. \qquad (9.13)$$

Equation (9.9) is the one-electron Schrödinger equation for an electron moving in the self-consistent potential field

$$V' = V + U + \hat{A}. \qquad (9.14)$$

Thus, the many-electron problem is reduced to the one-electron solution by neglecting the correlation between electrons due to their Coulomb repulsion. This requires additional justification, which will be given in § 12.

Hartree's and Fock's equations have been applied to various many-electron atoms and to multiatomic systems.

§10.   Heitler — London and Hund — Mulliken Methods

Two independent types of solution of Fock's equations are used for multiatomic systems: the Heitler—London (or atomic) method; and the Hund—Mulliken—Bloch (or molecular) method.

In the Heitler—London method, the one-electron functions ψ_i, which are used to construct a many-electron function of the system, are taken in the form of atomic functions, i.e., it is assumed that the electrons are distributed between atoms and ions and are localized near atomic cores. A solution of this type is exact if the atoms in a system are far apart and their electron wave functions overlap only a little.

In the Hund—Mulliken method, the one-electron wave functions extend over the whole system of atoms and have the same amplitudes for equivalent atoms. This means that each electron belongs to a system of atoms as a whole and moves independently of the other electrons. When applied to solids, the method is known as Bloch's method or model.

The Hund—Mulliken method is sometimes called the one-electron problem [4], and the Heitler—London method the many-electron approach. These names are valid only in the sense that in the Hund—Mulliken method we consider explicitly only one one-electron function and all the other functions are included in the expression for the potential, while in the Heitler—London method several one-electron functions are used explicitly. However, strictly speaking, both methods are one-electron approximations to the solution of the many-electron problem, because in both methods the total function of the system is made up from one-electron functions.

In the Hund—Mulliken method, the independent motion of each electron throughout a system means that there is no correlation in the distribution of electrons between the atoms, and although on the average all atoms may be neutral, virtual ionic states are formed when electrons aggregate at some atoms. The probability of virtual ionic states is of the same order of magnitude as the probability of neutral states. In the Heitler—London method, the correlation in the electron distribution between atoms is quite rigid — the electrons are distributed in equal numbers among various atoms, and there are no virtual ionic states. Since the electrons are located near atoms, the energy of electron repulsion decreases. The molecular-type wave functions are smoother than the atomic functions and, therefore, the average kinetic energy in the Heitler—London system is greater.

Thus, each of these methods is approximate and has its advantages and disadvantages. In their applications to relatively simple multiatomic systems, such as the hydrogen or lithium molecule, both methods give approximately the same results; the error in the determination of the binding energy is of the order of 0.5 eV per electron (p. 277 in [3]) and may be due to the omission of the correlation energy.

It is easily shown that, if we start with the atomic or molecular functions, the variational calculation according to Fock gives a self-consistent solution of the same type. We shall consider first the atomic-type functions. The Coulomb term

$$\sum_{j}{}' e^2 \int \frac{|\varphi_j(\zeta_2)|^2}{r_{12}} d\zeta_2$$

in Eq. (9.7), obtained from the Heitler—London function, screens part of the ionic potential of all the atoms except the i-th and, therefore, the solution of Eq. (9.7) is localized near the i-th atom, i.e., it may be a function of the atomic type. If the initial functions are of the molecular type, then the charges of all the ions are screened to the same degree, all the atoms are equivalent, and, therefore, the self-consistent solution may be of the molecular type.

Both these approximate methods are also widely used in solid-state theory, one or the other method being preferred depending on the nature of the problem. The Heitler–London model is more convenient for calculating binding forces, particularly in crystals with ionic and covalent bonds; it is used also in the theory of ferromagnetism. Bloch's model (corresponding to the Hund–Mulliken model) is more convenient in discussions of the electrical conductivity and other electrical properties of metals and semiconductors. The modern energy-band theory of solids is based mainly on Bloch's model.

The Heitler–London and Hund–Mulliken methods are equivalent in the case of electron configurations of the closed-shell type. In a crystal, this arises if we consider on its own a band filled completely with electrons; in this case, a determinant formed from the Heitler–London wave functions may be transformed into a determinant consisting of the Hund–Mulliken functions. Let ψ_i be a wave function localized near a certain atom; all ψ_i occur in the determinant in pairs for particles with opposite spins, so that the total Heitler–London function has the form:

$$\begin{vmatrix} \psi_1(\mathbf{r}_1)\,\eta_1(1) & \psi_1(\mathbf{r}_2)\,\eta_2(1) & \psi_1(\mathbf{r}_3)\,\eta_3(1)\cdots \\ \psi_1(\mathbf{r}_1)\,\eta_1(-1) & \psi_1(\mathbf{r}_2)\,\eta_2(-1) & \psi_1(\mathbf{r}_3)\,\eta_3(-1)\cdots \\ \psi_2(\mathbf{r}_1)\,\eta_1(1) & \psi_2(\mathbf{r}_2)\,\eta_2(1) & \psi_2(\mathbf{r}_3)\,\eta_3(1)\cdots \\ \psi_2(\mathbf{r}_1)\,\eta_1(-1) & \psi_2(\mathbf{r}_2)\,\eta_2(-1) & \psi_2(\mathbf{r}_3)\,\eta_3(-1)\cdots \\ \cdots\cdots\cdots & \cdots\cdots\cdots & \cdots\cdots\cdots \end{vmatrix} \quad (10.1)$$

Here, the elements of all the odd rows have the same spin functions, and the same is true for all the even rows. We shall add to each of the odd rows a linear combination of the remaining odd rows and we shall do the same for the even rows. As a result of this operation, we obtain the determinant

$$\begin{vmatrix} \psi_{\mathrm{I}}(\mathbf{r}_1)\,\eta_1(1) & \psi_{\mathrm{I}}(\mathbf{r}_2)\,\eta_2(1) & \psi_{\mathrm{I}}(\mathbf{r}_3)\,\eta_3(1)\cdots \\ \psi_{\mathrm{I}}(\mathbf{r}_1)\,\eta_1(-1) & \psi_{\mathrm{I}}(\mathbf{r}_2)\,\eta_2(-1) & \psi_{\mathrm{I}}(\mathbf{r}_3)\,\eta_3(-1)\cdots \\ \psi_{\mathrm{II}}(\mathbf{r}_1)\,\eta_1(1) & \psi_{\mathrm{II}}(\mathbf{r}_2)\,\eta_2(1) & \psi_{\mathrm{II}}(\mathbf{r}_3)\,\eta_3(1)\cdots \\ \psi_{\mathrm{II}}(\mathbf{r}_1)\,\eta_1(-1) & \psi_{\mathrm{II}}(\mathbf{r}_2)\,\eta_2(-1) & \psi_{\mathrm{II}}(\mathbf{r}_3)\,\eta_3(-1)\cdots \\ \cdots\cdots\cdots & \cdots\cdots\cdots & \cdots\cdots\cdots \end{vmatrix} \quad (10.2)$$

where ψ_{I}, ψ_{II}, etc., are various independent linear combinations of the atomic functions ψ_i. These linear combinations can be selected so that they are Hund–Mulliken wave functions. However, the determinants (10.1) and (10.2) are identical because (10.2) is obtained from (10.1) by combining various rows. Thus, in this case, the Heitler–London and Hund–Mulliken methods are fully equivalent. The proof, of course, is valid on condition that the number of the wave functions ψ_i is equal to the number of electron pairs, i.e., it is valid for closed shells or filled bands.

An attempt to unify the Heitler–London and Hund–Mulliken methods and thus improve the accuracy has been made by Schmid [5], who calculated the binding energy in the diamond lattice. He considered a crystal of diamond to be, in the first approximation, an assembly of valence bonds each of which was formed by two electrons; the many-electron problem was thus reduced to the two-electron problem. The expression for a two-electron wave function Schmid wrote in the form:

$$\varphi_{ij}(\zeta_1,\ \zeta_2) = \frac{1}{\sqrt{2\,(1+A^2)}}\ \{[\psi_i(\mathbf{r}_1)\,\psi_j(\mathbf{r}_2) + \psi_j(\mathbf{r}_1)\,\psi_i(\mathbf{r}_2)] +$$

$$+ A\,[\psi_i(\mathbf{r}_1)\,\psi_i(\mathbf{r}_2) + \psi_j(\mathbf{r}_1)\,\psi_j(\mathbf{r}_2)]\}\ \frac{\eta_1(s_1)\,\eta_2(s_2) - \eta_2(s_1)\,\eta_1(s_2)}{\sqrt{2}}, \qquad (10.3)$$

where ψ are functions of coordinates and η are functions of spin.

The first square brackets in Eq. (10.3) give the Heitler–London function, and the second give the virtual ionic-state functions. The parameter A is found from the condition for minimum energy of the whole crystal, which is calculated from the formula

$$E = \frac{\int \Psi \hat{H} \Psi\, d\zeta}{\int |\Psi|^2\, d\zeta}, \qquad (10.4)$$

where Ψ is the total wave function of the crystal, which is in the form of a determinant consisting of functions like (10.3). If $A = 0$, Eq. (10.3) transforms into a Heitler–London function, and if $A = 1$, it becomes a Hund–Mulliken function because when $A = 1$ the coordinate part of the function may be rewritten in the form:

$$[\psi_i(\mathbf{r}_1) + \psi_j(\mathbf{r}_1)]\,[\psi_i(\mathbf{r}_2) + \psi_j(\mathbf{r}_2)]. \qquad (10.5)$$

A calculation gave the value $A = 0.85$, i.e., the best function was in this case close to the Hund–Mulliken function. The function (10.3) may also be represented as a linear combination of a Heitler–London function φ_{HL} and a Hund–Mulliken function φ_{HM}, the energy minimum occurring when

$$\varphi = 0.15\varphi_{HL} + 0.85\varphi_{HM}. \qquad (10.6)$$

§ 11. Method of Elementary Excitations

The method of elementary excitations, developed mainly by Vonsovskii and his school [6-13] and extended in the work of Bogolyubov and Tyablikov [14], is based on the Heitler—London method as applied to crystals, and allows one to include ionic and other excited states. Initially, the method was used in the theory of ferromagnetism, but later it was employed to explain the electrical conductivity and other properties of metals and semiconductors.

The essence of the method is to assume that in the ground state of a system there is one electron per atom; the electrons are described, as in the Heitler—London method, by atomic functions of the atomic ground state, and all the electrons have spins directed in the same way. The latter is true of ferromagnets. In the case of semiconductors, it is more logical to assume that there are two electrons per atom and that the spins in each pair are opposite. The ground state is not calculated but only deviations from it are considered, and these are called elementary excitations, i.e., changes in the quantum-mechanical states of single electrons, which tend to increase the energy of the whole system. These excitations may be transferred from atom to atom, they may be propagated in the form of waves along the crystal, they may have a definite effective mass, and they may carry a charge, i.e., they behave as quasiparticles.

The main types of elementary excitation are the following:

1) a pair of electrons near an atom, but if the ground state has two electrons (semiconductors) the pair is replaced by three electrons. In accordance with the Pauli principle, the spins of the electrons in a pair should be opposite, but in the case of three electrons, one electron should be in an excited state; since an electron pair or trio has an excess electron, it carries a negative charge equal to one electronic charge;

2) a hole, which represents an absence of an electron near an atom; a hole has a single positive charge on a "background" of atoms in the ground state;

3) a ferromagnon, which is an electron in the ground state but with a reversed spin, i.e., with right-handed spin;

4) a left-handed exciton, which is an electron in the excited state of an atom but with "normal" left-handed spin;

5) a right-handed exciton, which is also an electron in the excited state but with the spin reversed.

Since the charge of ferromagnons and excitons is the same as that of normal atoms, these quasiparticles do not carry charge and do not take part in the passage of a current but they do transport energy.

Apart from those listed, combined elementary excitations are possible: for example, a pair in which one of the electrons is in the excited state and has right-handed or left-handed spin; however, the excitation energy of such quasiparticles is obviously higher than those listed above and, therefore, their number will be small. Moreover, we can, in principle, discuss not one but many excited states of each atom and then the number of types of exciton will increase accordingly.

If all the various types of excitation are included, the problem cannot be solved by normal means and the second quantization method must be used. In the second quantization method, the independent variables are not the electron coordinates but the electron population numbers of various states, i.e., in our case, the numbers of various quasiparticles. The second quantization operators act on the population numbers, and change them, i.e., these operators are quasiparticle creation and annihilation operators.

The Hamiltonian (9.1) for many-electron systems has the following form in the second quantization representation:

$$\hat{H} = U_0 + \sum_{n,\,\lambda} L_{n\lambda n'\lambda'} \hat{a}_{n\lambda}^+ \hat{a}_{n'\lambda'} +$$

$$+ \frac{1}{2} \sum_{n_1,\,\lambda_1,\,n_2,\,\lambda_2,\,n_1',\,\lambda_1',\,n_2',\,\lambda_2'}' F_{n_1\lambda_1 n_2\lambda_2 n_1'\lambda_1' n_2'\lambda_2'} \hat{a}_{n_1\lambda_1}^+ \hat{a}_{n_1'\lambda_1'}^+ \hat{a}_{n_1\lambda_1} \hat{a}_{n_1'\lambda_1'};$$

(11.1)

here U_0 is the additive part of the operator; n is the number of atoms, i.e., lattice sites; the index λ describes the state of an electron of a given atom, including the direction of its spin; $\hat{a}_{n\lambda}^+$ and $\hat{a}_{n\lambda}$ are the electron emission and absorption operators for an atom in the state λ; and L and F with their indices are matrix elements defined by the formulas

$$L_{n\lambda n'\lambda'} = \int \psi_{n\lambda}^*(\mathbf{r}) \left[-\frac{\hbar^2}{2m} \nabla^2 + U_0 \right] \psi_{n'\lambda'}(\mathbf{r})\, d\mathbf{r},$$

(11.2)

$$F_{n_1\lambda_1 n_2\lambda_2 n_1'\lambda_1' n_2'\lambda_2'} = e^2 \int\int \frac{\psi_{n_1\lambda_1}^*(\mathbf{r}') \psi_{n_2\lambda_2}(\mathbf{r}') \psi_{n_1'\lambda_1'}(\mathbf{r}) \psi_{n_2'\lambda_2'}(\mathbf{r})\, d\mathbf{r}d\mathbf{r}'}{|\mathbf{r} - \mathbf{r}'|},$$

(11.3)

where $\psi_{n\lambda}(\mathbf{r})$ are the atomic wave functions.

The first sum in (11.1), which contains paired products of the second quantization operators, gives the kinetic energy of the electrons and their potential energy in a given field (the ionic field), while the pair interactions between electrons are described by quartic products of the operators in the second sum.

The expression (11.1) is an exact Hamiltonian of a many-electron system and an exact many-electron solution would be obtained by solving Schrödinger's equation with this Hamiltonian using a sufficient number of excited atomic states. However, it has not yet been possible to diagonalize the sum of quartic products of the operators and they are approximately replaced by paired products; thus the interaction between the electrons is replaced by an effective field, as in the self-consistent field method.

In order to reduce the quartic terms in (11.1) to quadratic terms the following approximations must be made.

1. It is assumed that the number of elementary excitations is small, i.e., the average population number of the ground state is close to unity, and of the other states is much less than unity. However, the population number of a state of the n-th atom is equal to $\hat{a}_{n\lambda}^+ \hat{a}_{n\lambda}$, and, consequently, assuming that $\lambda = 0$ for the ground state, we have on the average

$$\hat{a}_{n0}^+ \hat{a}_{n0} \approx 1; \quad \hat{a}_{n\lambda}^+ \hat{a}_{n\lambda} \ll 1 \quad \text{if} \quad \lambda \neq 0. \tag{11.4}$$

2. Averaging over the background is carried out, i.e., the products of the operators $\hat{a}_{n0}^+ \hat{a}_{n0}$ are replaced by their average values, which are approximately unity. In this way, we ignore the correlation in the distribution of the electrons over states even for neighboring atoms. Averaging over the background is equivalent to averaging the electron charge in the self-consistent field method.

3. The quartic terms in which $\lambda = 0$ is not repeated twice are ignored [because of the inequality of Eq. (11.4)]. Consequently, if only the electron excitations listed earlier are taken into account, the Hamiltonian, after some transformations, becomes

$$\hat{H} = U_0 + \Delta E (s_4 + s_5) + s_1 A + \sum_{n,\,n'} J_{nn'} (\hat{a}_{n'3} \hat{a}_{n3}^+ - \hat{a}_{n3} \hat{a}_{n'3}^+) +$$
$$+ \sum_{n,\,n'} J_{nn'}^* (\hat{a}_{n5} \hat{a}_{n'5}^+ - \hat{a}_{n5} \hat{a}_{n'5}^+) + \sum_{n,\,n'} \beta_{nn'} (\hat{a}_{n1} \hat{a}_{n'1}^+ - \hat{a}_{n2} \hat{a}_{n'2}^+); \tag{11.5}$$

here s_1, s_2, s_3, s_4, and s_5 are the numbers of pairs, trios, ferromagnons, left-handed and right-handed excitons, respectively; $\hat{a}$ and $\hat{a}^+$, with the corresponding indices, are their creation and annihilation operators; ΔE is the difference between the excited and ground state energies of an atom; A is the electrostatic repulsion energy for electrons in a pair; $J_{nn'}$ is the exchange integral for electrons with different spins in the nonexcited state; $J_{nn'}^*$ is the exchange integral for electrons in the ground and excited states; and $\beta_{nn'}$ is the transport integral for a pair.

The Hamiltonian (11.5) is easily diagonalized by expanding the operators $\hat{a}$ as a Fourier series. This gives waves propagated along the whole crystal, as in Bloch's model. A many-electron system reduces to individual quasiparticles described by one-particle equations. The main achievement of this method has been to describe excitations not involving currents: ferromagnons and excitons. Pairs and trios do not differ essentially from electrons and holes in the normal band theory if we do not look too closely at the details of the calculations of all the possible exchange integrals. Actual calculations carried out by the elementary excitation method, for example, for the valence band of diamond-type lattices [15], yield exactly the same results as the usual one-electron approximation [16].

The many-electron nature of the solution by means of the elementary excitation method is frequently stressed. In fact, the many-electron approach is restricted only to the initial equations and is lost when the quartic products of the second quantization operators are replaced by quadratic products; the method is one of the variants of the one-electron approximation, like the Heitler—London method.

The disadvantage of the method of elementary excitations compared with Bloch's model is the greater complexity of the calculations; therefore, this method has not been used to obtain the same number of actual results as Bloch's model.

In spite of these criticisms, the value of the elementary excitation method for the theory of the electrical conductivity of solids lies in the fact that the one-electron approximation is obtained in a way different from that used in Bloch's model and, consequently, although the method of elementary excitations is not a rigorous proof of the validity of the one-electron approximation, it is at least additional evidence of this validity.

§12. Justification of the One-Electron Approximation by the Theory of Collective Electron Interaction

The validity of the one-electron approximation may be proved and the domain of its validity determined only by means of a theory which does not neglect the correlation in the motion of the electrons, i.e., in which, in addition to the motion of individual electrons, their collective movement is considered. This has been done in a very simple and lucid way by Bohm and Pines [17-19], whose treatment will now be summarized.

Bohm and Pines used a simplified model of a metal in the form of a gas of point-like electrons moving in a field of uniformly distributed positive charge, i.e., the crystal structure and space periodicity are neglected. Under the action of Coulomb forces, the mobile electrons tend to screen any charge e introduced into the system, but the thermal motion of the electrons prevents complete screening and, therefore, an equilibrium distribution of electrons is established in which, as is known from the Debye—Hückel theory of electrolytes [20], the potential of a screened point charge q is given by the formula

$$U = \frac{q}{r} e^{-\frac{r}{\lambda_D}}, \qquad (12.1)$$

where r is the distance from the charge, and λ_D is the Debye screening radius

$$\lambda_D = \sqrt{\frac{k_B T}{4\pi n_0 e^2}}, \qquad (12.2)$$

k_B is Boltzmann's constant, T is the absolute temperature, and n_0 is the average electron density.

The screening of the charges means that at distances greater than λ_D we need not consider the motion of individual electrons but their collective motion in the form of plasma oscillations. Conversely, at distances shorter than λ_D the electrons move independently of one another. To establish a relationship between the collective and individual motions of the electrons we must introduce a concept of charge density fluctuations

$$\rho_k = \sum_i e^{-i\mathbf{k} \cdot \mathbf{r}_i}, \qquad (12.3)$$

where the summation is carried out for all electrons with radius vectors $\mathbf{r}_i$. A fluctuation may be represented as a sum of two terms:

$$\rho_{\mathbf{k}} = q_{\mathbf{k}} + \eta_{\mathbf{k}}, \tag{12.4}$$

of which $q_{\mathbf{k}}$ describes the collective motion of the electron plasma, and $\eta_{\mathbf{k}}$ gives the motion of individual electrons.

It is easily shown that if $k > 1/\lambda_D$ then $\eta_{\mathbf{k}}$ predominates, but if $k < 1/\lambda_D$ then the term $q_{\mathbf{k}}$, describing the plasma oscillations, is the dominant one. If the thermal motion of the electrons is ignored the frequency of the plasma oscillations is given by

$$\omega_p = \sqrt{\frac{4\pi n_0 e^2}{m}} \tag{12.5}$$

(m is the electronic mass, n_0 is the electron density), but if the thermal motion is included this frequency is found from the equation

$$1 = \frac{4\pi n e^2}{m} \sum_i \frac{1}{(\omega - \mathbf{k} \cdot \mathbf{v}_i)^2}, \tag{12.6}$$

(where $\mathbf{v}_i$ is the velocity of the i-th electron).

The initial Hamiltonian of the system, expressed in terms of $\rho_{\mathbf{k}}$, can be conveniently written in the form:

$$\hat{H}_0 = \sum_i \frac{p_i^2}{2m} + 2\pi e^2 \sum_{\mathbf{k}} \frac{\rho_{\mathbf{k}}^* \rho_{\mathbf{k}}}{k^2} - 2\pi n e^2 \sum_{\mathbf{k}} \frac{1}{k^2}; \tag{12.7}$$

here the second term includes the energy of electron interaction, and the third term subtracts this energy. $\hat{H}_0$ is expressed completely in terms of the coordinates of individual electrons. In order to describe the collective motion of electrons, we introduce new additional coordinates $Q_{\mathbf{k}}$ and corresponding momenta $P_{\mathbf{k}}$. Consequently, the Hamiltonian acquires additional terms

$$\hat{H}_{add} = \sum_{k < k_c} \frac{\hat{P}_{\mathbf{k}}^* \hat{P}_{\mathbf{k}}}{2} - i \sum_{k < k_c} \sqrt{\frac{4\pi e^2}{k^2}} \, P_{\mathbf{k}} \rho_{\mathbf{k}}. \tag{12.8}$$

The number of the collective degrees of freedom is

$$n' = \frac{k_c^3}{6\pi^2}. \tag{12.9}$$

In order to retain the total number of degrees of freedom of the system, the wave function must satisfy the same number of additional conditions of the form

$$P_k \psi = 0 \quad (k < k_c). \tag{12.10}$$

For a complete system of electron and collective coordinates, with $3n + n'$ degrees of freedom, the Hamiltonian is the sum of the expressions (12.7) and (12.8) and may be written in the form:

$$\hat{H} = \sum_i \frac{p_i^2}{2m} + \frac{1}{2} \sum_{k < k_c} \left(P_k^* - i\rho_k \sqrt{\frac{4\pi e^2}{k^2}} \right) \left(P_k + i\rho_k^* \sqrt{\frac{4\pi e^2}{k^2}} \right) -$$
$$- 2\pi n e^2 \sum_{k < k_c} \frac{1}{k^2} + 2\pi e^2 \sum_{\substack{i \neq j \\ k > k_c}} \frac{1}{k^2} e^{i\mathbf{k} \cdot (\mathbf{r}_i - \mathbf{r}_j)}. \tag{12.11}$$

If, in solving Schrödinger's equations, we take into account the conditions in Eq. (12.10), we again obtain the Hamiltonian of Eq. (12.7). In order to obtain collective excitations, we shall solve Schrödinger's equation

$$\hat{H}\psi = E\psi, \tag{12.12}$$

without satisfying exactly the auxiliary conditions (12.10). This is approximately justified by the fact that the number of these auxiliary conditions is small: $n' \ll 3n$.

In order to relate P_k and ρ_k, we canonically transform the wave function

$$\psi = e^{\frac{iS}{\hbar}} \Phi, \tag{12.13}$$

where

$$S = -i \sum_{k < k_c} \sqrt{\frac{4\pi e^2}{k^2}} Q_k \rho_k^*. \tag{12.14}$$

From this transformation, we obtain a new Schrödinger equation

$$\hat{H}_c \Phi = E\Phi, \tag{12.15}$$

where $\hat{H}_c$ is the initial collective Hamiltonian:

$$\hat{H} = \sum_i \frac{p_i^2}{2m} + \sum_{k < k_c} \frac{P_k^* P_k + \omega_p^2 Q_k^* Q_k}{2} +$$

$$+ \sum_{i, k < k_c} \frac{\sqrt{4\pi e^2}}{m} \mathbf{i_k} \left(\mathbf{p}_i - \frac{\hbar k}{2}\right) Q_k e^{i\mathbf{k} \cdot \mathbf{r}_i} + 2\pi e^2 \sum_{\substack{i \neq j \\ k > k_c}} \frac{e^{i\mathbf{k} \cdot (\mathbf{r}_i - \mathbf{r}_j)}}{k^2} -$$

$$- 2\pi n e^2 \sum_{k < k_c} \frac{1}{k^2} + \frac{2\pi e^2}{m} \sum_{\substack{i, k, l < k_c \\ k \neq -l}} \mathbf{i_k} \mathbf{i_l} Q_k Q_l e^{i(\mathbf{k}+\mathbf{l}) \cdot \mathbf{r}_i}; \quad (12.16)$$

$\mathbf{i_k}$ and $\mathbf{i_l}$ are unit vectors directed along the plasma oscillations.

The long-range part of the Coulomb potential is omitted completely from Eq. (12.16). The first three terms describe the kinetic energy of the electrons, the energy of the plasma oscillations, and the interaction of the electrons with the plasma. This interaction is represented by an effective coupling constant

$$g^2 = \left(\frac{\overline{\mathbf{k} \cdot \mathbf{p}_i}}{m\omega_p}\right)^2 \simeq \frac{\beta^2}{2r_1}, \quad (12.17)$$

where the bar represents averaging over the electron momenta $\beta = k_c/k_F$; $\mathbf{k}_F$ is the wave vector of an electron near a Fermi level; and r_1 is the average distance between electrons, measured in Bohr radii. Numerical estimates show that for real metals $g^2 \ll 1$.

The fourth term in Eq. (12.16) describes the short-range electron interaction, which we shall denote by $\hat{H}_{sr}$, and the last term represents complex interaction between the plasma and the electrons, which is small and can always be neglected.

The additional conditions of Eq. (12.10) now become:

$$\left(P_k - i\rho_k \sqrt{\frac{4\pi e^2}{k^2}}\right) \Phi = 0. \quad (12.18)$$

So far, the collective motion is separated out incompletely because there is still weak interaction of the plasma with the electrons and the coupling of the plasma oscillations with the motion of the electrons through the conditions given by Eq. (12.18). However, both these effects may be eliminated to within g^2 by the usual canonical transformation of perturbation theory, after which the Hamiltonian of Eq. (12.16) becomes:

$$\hat{H} = \sum_i \frac{p_i^2}{2m}\left(1 - \frac{\beta^3}{6}\right) + \sum_{k < k_c} \left(\frac{P_k^* P_k + \omega^2 Q_k^* Q_k}{2} - \frac{2\pi n e^2}{k^2}\right) + \hat{H}_{sr} -$$

$$-\frac{\pi e^2}{m^2} \sum_{\substack{k < k_c \\ i \neq j}} \frac{\left[\mathbf{i_k}\left(\mathbf{p}_i - \frac{\hbar \mathbf{k}}{2}\right)\right]\left[\mathbf{i_k}\left(\mathbf{p}_j + \frac{\hbar \mathbf{k}}{2}\right)\right]}{\omega\left(\omega - \frac{\mathbf{k}\mathbf{p}_j}{m} - \frac{\hbar^2}{2m}\right)} e^{i\mathbf{k}\cdot(\mathbf{r}_i - \mathbf{r}_j)} +$$

(12.19)

$+$ complex conjugates;

ω is the frequency of independent collective oscillations, which is given by the dispersion relationship

$$1 = \frac{4\pi e^2}{m} \sum_i \frac{1}{\left(\omega - \frac{\mathbf{k}\cdot\mathbf{p}_i}{m}\right)^2 - \frac{\hbar^2 k^4}{4m^2}}.$$

(12.20)

The interaction of the plasma with the electrons is unimportant and reduces to an increase of the effective electron mass, which becomes

$$m_{\text{eff}} = \frac{m}{1 - \frac{\beta^3}{6}},$$

(12.21)

and to a slight change in the dispersion relationship for the plasma oscillations: Eq. (12.20) instead of Eq. (12.6). The last term in Eq. (12.19) is small. Thus the Hamiltonian of Eq. (12.16) or Eq. (12.19) describes almost independent electrons.

When the interaction of electrons with the plasma and with other electrons is small, the wave function of the system may be represented in the form

$$\Phi = \psi_{\text{osc}} \chi_0;$$

(12.22)

here ψ_{osc} is the product of the functions of harmonic oscillators describing the ground states of the plasma, and χ_0 is a determinant composed of the wave functions of free electrons.

Ignoring small terms in the Hamiltonian (12.16) and applying it to the function in Eq. (12.22), we obtain the ground-state energy of the system in the zeroth approximation:

$$E = \frac{3}{5} n E_0 + \sum_{k < k_c} \left(\frac{\hbar \omega_p}{2} - \frac{2\pi n e^2}{k^2}\right) + \bar{H}_{\text{sr}}.$$

(12.23)

The Coulomb correlation energy, i.e., the gain in the energy because electrons tend to repel one another by Coulomb forces, is equal to the dif-

ference between the zero-point plasma oscillations and the energy of interaction of electrons calculated in the Hartree−Fock approximation. Calculations show that the correlation energy per electron is, in the zeroth
approximation,

$$E^0_{corr} = \frac{0.866\beta^3}{r_1^{3/2}} - \frac{0.458\beta^2}{r_1} + 0.019\frac{\beta^4}{r_1} \text{ rydbergs.} \qquad (12.24)$$

The quantity β, i.e., the threshold value of the wave number for the
plasma oscillations, is found from the condition for the correlation energy
minimum. If the last small term is neglected in Eq. (12.24), we find that

$$\beta = 0.353\sqrt{r_1}; \quad E^0_{corr} = -(0.019 - 0.0003r_1) \text{ rydbergs.} (12.25)$$

In the next approximation, the Hamiltonian of Eq. (12.16) includes the
terms describing the interaction between the plasma and the electrons and the
short-range interaction between electrons themselves. Therefore, we can calculate the correction to the correlation energy due to these terms. If corrections are added to Eq. (12.25), the total correlation energy becomes

$$E_{corr} = -(0.114 - 0.0313 \ln r_1 + 0.0005r_1) \text{ rydbergs.} \qquad (12.26)$$

This value should be compared with the average kinetic energy of the
electrons, which is $E_{kin} = 2.21/r_1^2$. In metals with a high electron density and
$r_1 \simeq 2$, the correlation energy is several times smaller than the kinetic energy,
and this justifies the use of the perturbation theory. The situation is less fav-
orable in the case of such metals as cesium, for which $r_1 = 5.5$ and the ratio
of the two energies approaches unity, i.e., calculations using the perturbation
theory are at the limits of their validity.

Thus a system of electrons in a metal may be represented in terms of
nearly independent electrons, mutually interacting by relatively weak short-
range forces. The long-range part of the Coulomb interaction is described
by plasma oscillations. The frequency of these oscillations is quite high: for
metals, it is − according to Eq. (12.5) − of the order of 10^{16} cps. The ratio
of the energy of a quantum of plasma oscillations and the energy of an electron at a Fermi level is

$$\frac{\hbar\omega_p}{E_0} = \sqrt{\frac{16\pi n_0 e^2 \hbar^2}{m^3 v_0^4}} \approx \sqrt{r_1}. \qquad (12.27)$$

Since for all metals $r_1 > 1$, the above ratio is always greater than unity.
Consequently, none of the electrons in a metal may excite plasma oscillations.

Thus the independent electron model or the one-electron approximation is justified so long as we are dealing with intrinsic electrons in a metal or with electrons of low energies. High-energy electrons (for example, β rays) absorbed by a metal are capable of exciting plasma oscillations, and their motion cannot be considered in the one-electron approximation. The limit of applicability of the one-electron approximation expressed in terms of the electron energy is $\hbar\omega_p$.

All calculations in the present section have been carried out for a smoothed-out potential. We may, however, expect that the superposition of a periodic potential does not alter essentially the interaction of electrons with each other and the correlation in their motion, so that the conclusions of the theory of collective electron interaction are applicable to real metals and to filled bands in semiconductors.

§13. Electron in a Periodic Field

Having justified the one-electron approximation, we shall now use it, i.e., we shall consider the motion of a single electron in some self-consistent field V. This motion is described by Schrödinger's equation

$$-\frac{\hbar^2}{2m}\nabla^2\psi + V\psi = E\psi. \tag{13.1}$$

Irrespective of the actual nature of the potential V and the method of finding it for a crystal with a regular distribution of atoms, we can always say that this potential has spatial periodicity, i.e., that

$$V(\mathbf{r}) = V(\mathbf{r} + \mathbf{R}_j), \tag{13.2}$$

where $\mathbf{R}_j$ is the translation vector of the crystal lattice,

$$\mathbf{R}_j = n_1\mathbf{a}_1 + n_2\mathbf{a}_2 + n_3\mathbf{a}_3, \tag{13.3}$$

which brings the crystal into self-coincidence. Here, $\mathbf{a}_1$, $\mathbf{a}_2$, $\mathbf{a}_3$ are the vectors of elementary translations in the crystal; n_1, n_2, and n_3 are integers.

We shall introduce symmetry operators representing the translation group of the crystal. An element $\hat{T}_j$ of this group translates the crystal by a vector $\mathbf{R}_j$, i.e.,

$$\hat{T}_j\mathbf{r} = \mathbf{r} + \mathbf{R}_j. \tag{13.4}$$

It follows from Eq. (13.2) that $V(\mathbf{r})$ is invariant with respect to any operator $\hat{T}_j$. This means that the operators $\hat{T}_j$ commute with the Hamiltonian $\hat{H}$, and from Eq. (13.4) it follows that they commute with each other. Therefore, the solutions $\psi_{\mathbf{k}}$ of Schrödinger's equation (13.1) are also eigenfunctions of the operators $\hat{T}_j$, in other words,

$$\hat{T}_j \psi_{\mathbf{k}} = \sigma_{j\mathbf{k}} \psi_{\mathbf{k}}. \qquad (13.5)$$

The eigenvalue of the operator $\hat{T}_j - \sigma_{j\mathbf{k}}$ can be conveniently represented in the form $\sigma_{j\mathbf{k}} = e^{i\mathbf{k} \cdot \mathbf{R}_j}$, and, therefore, the application of two translations $\hat{T}_j$ and $\hat{T}_{j'}$ multiplies the wave function by $\sigma_{j\mathbf{k}}\sigma_{j\mathbf{k}} = e^{i\mathbf{k} \cdot (\mathbf{R}_j + \mathbf{R}_{j'})}$, and Eq. (13.5) becomes:

$$\psi_{\mathbf{k}}(\mathbf{r} + \mathbf{R}_j) = e^{i\mathbf{k} \cdot \mathbf{R}_j}\psi(\mathbf{r}). \qquad (13.6)$$

We shall introduce a new function

$$u_{\mathbf{k}}(\mathbf{r}) = e^{-i\mathbf{k} \cdot \mathbf{r}}\psi_{\mathbf{k}}(\mathbf{r}). \qquad (13.7)$$

It follows from Eq. (13.6) that

$$u_{\mathbf{k}}(\mathbf{r} + \mathbf{R}_j) = u_{\mathbf{k}}(\mathbf{r}). \qquad (13.8)$$

It means that the solution of Eq. (13.1) for a crystal has the form

$$\psi_{\mathbf{k}}(\mathbf{r}) = e^{i\mathbf{k} \cdot \mathbf{r}}u_{\mathbf{k}}(\mathbf{r}), \qquad (13.9)$$

where the function $u_{\mathbf{k}}(\mathbf{r})$ has exactly the same periodicity as the lattice. The functions (13.9) were first proposed by Bloch [21] and are known as Bloch waves; the whole proof is known as Bloch's theory, and the relationships in Eq. (13.6) are known as Bloch's conditions.

In the one-dimensional case, the expressions (13.8) and (13.9) become:

$$\psi(x) = e^{ikx}u(x); \quad u(x + a) = u(x). \qquad (13.10)$$

A purely mathematical proof of the fact that Eq. (13.10) is a solution of a differential equation with a periodic coefficient, was known much before Bloch as Floquet's theorem.

So far, we have imposed no restrictions on the vector $\mathbf{k}$, and, in general, it may be complex. However, if $\mathbf{k}$ is complex, then in an infinite crys-

tal lattice the wave function (13.9) may rise without limit along some directions, which is physically meaningless, and such lattices must be rejected.
Real solutions for an infinite lattice contain a purely real $\mathbf{k}$. For real $\mathbf{k}$,
Bloch's function (13.9) is a plane wave modulated periodically by a multi-
plier $u_{\mathbf{k}}$. Because of the periodicity, the probability of finding an electron
is the same for all unit cells in a crystal, i.e., the wave function (13.9) applies to the whole crystal and describes the quasifree motion of an electron
over the whole crystal; the periodically distributed atoms do not scatter electrons and this gives rise to the high electrical conductivity of metals.

Whether $\mathbf{k}$ is complex or real depends only on the value of the energy
$E_{\mathbf{k}}$ which occurs in Eq. (13.1). The allowed energy values are those which
correspond to real value of $\mathbf{k}$, and the forbidden ones are those which correspond to complex $\mathbf{k}$. In the one-dimensional case, it can be shown mathematically that when $E_{\mathbf{k}}$ is varied from minus infinity to plus infinity, we have
a succession of allowed and forbidden energy bands, and the energy spectrum
of electrons in a periodic field has a band structure. This mathematical proof
is known as Mathieu's problem.

Slater [22, 23] proved the same for a three-dimensional crystal by reducing it to the one-dimensional case. For this purpose, the potential V is
expanded in a triple Fourier series:

$$V(\mathbf{r}) = \sum_{\mathbf{k}_j} V(\mathbf{k}_j) e^{i\mathbf{k}_j \cdot \mathbf{r}}. \tag{13.11}$$

We shall assume that the main terms of the series are of the form:

$$\pm k_1 = \frac{2\pi}{a_1}(\pm 1,\ 0,\ 0), \quad \pm k_2 = \frac{2\pi}{a_2}(0,\ \pm 1,\ 0), \quad \pm k_3 = \frac{2\pi}{a_3}(0,\ 0,\ \pm 1),$$

and all the remaining terms can be neglected.

Then the potential of Eq. (13.11) becomes:

$$V(\mathbf{r}) = 2 \sum_{\alpha=1}^{3} V_\alpha \left(1 - \cos \frac{2\pi x}{a_\alpha}\right) \tag{13.12}$$

(V_α are constants) and the variables can be separated if we assume that

$$E = \sum_{\alpha=1}^{3} E_\alpha \quad \text{and} \quad \psi = \prod_{\alpha=1}^{3} \psi_\alpha(x_\alpha). \tag{13.13}$$

The substitution of Eq. (13.13) into Eq. (13.1) leads to the one-dimensional Schrödinger equation

$$-\frac{\hbar^2}{2m} \cdot \frac{d^2 \psi_2}{dx^2} + 2V_\alpha \left(1 - \cos \frac{2\pi x}{a_\alpha}\right) = E_\alpha \psi_\alpha(x). \qquad (13.14)$$

The substitutions

$$w = \frac{\pi x}{a}, \quad s = \frac{8ma^2 V_\alpha}{\pi^2 \hbar^2}, \quad \varepsilon_\alpha = \sqrt{\frac{2m}{V_\alpha}} \cdot \frac{a E_\alpha}{2\pi \hbar} \qquad (13.15)$$

reduce Eq. (13.14) to the dimensionless equation

$$-\frac{d^2 \psi}{dw^2} + \frac{1}{2} s (1 - \cos 2w)\, \psi = \varepsilon_\alpha \sqrt{s}\,\, \psi, \qquad (13.16)$$

which is the well-known Mathieu equation for which the spectrum of the eigenvalues ε_α is known: it consists of a series of allowed bands, separated by forbidden bands.

Returning to the three-dimensional problem, we note that the energy is the sum of three terms, representing motion in three directions: $\varepsilon = \varepsilon_1 + \varepsilon_2 + \varepsilon_3$. The allowed values of the total energy split into bands but these bands overlap at low interatomic distances. Moreover, in order to approach the real crystal case, we shall use as perturbations the terms rejected in expanding Eq. (13.11). This gives rise to the splitting of some levels, but the band structure of the spectrum is retained. Details of the three-dimensional calculations are given in Slater's work [22].

§14. Principal Methods of Calculating the Energy Spectrum of Electrons in a Crystal

The discussion presented in the preceding section has established the general nature of the wave functions and the energy spectrum of electrons in a crystal in general, but has left unsolved the problem of the quantitative calculation for real crystals, in particular the determination of the structure of the energy bands, of their width, and of the dependence $E(\mathbf{k})$.

At present, there are many methods which claim to give quantitative results; they are reviewed in [23]. A complete review is impossible here so that we shall touch briefly on the most widely used methods, i.e., the method of tightly bound electrons, the method of cells, the method of orthogonalized plane waves, and the equivalent orbital method.

In the first method, electrons are regarded as being tightly bound to atoms and in the zeroth approximation the wave function of a crystal is a linear combination of the atomic functions χ_n, and it is in the form of a Bloch sum

$$\psi_k(\mathbf{r}) = \frac{1}{\sqrt{G}} \sum_n e^{i\mathbf{k}\cdot\mathbf{R}_n}\chi(\mathbf{r} - \mathbf{R}_n), \tag{14.1}$$

which satisfies Bloch's conditions (13.6); G is the number of atoms in that part of the crystal under consideration.

The n-th atomic function also satisfies Schrödinger's equation for the n-th atom,

$$-\frac{\hbar^2}{2m}\nabla^2\chi(\mathbf{r} - \mathbf{R}_n) + U(\mathbf{r} - \mathbf{R}_n)\,\chi(\mathbf{r} - \mathbf{R}_n), = E^0\chi(\mathbf{r} - \mathbf{R}_n), \tag{14.2}$$

where $U(\mathbf{r} - \mathbf{R}_n)$ is the atomic potential.

Substituting Eq. (14.1) into Eq. (13.1) (the equation for a crystal), multiplying it by $\chi^*(\mathbf{r} - \mathbf{R}_n)$, integrating over the whole volume, and neglecting overlap integrals of various atomic functions, we obtain an expression for the energy

$$E_k = E^0 - \alpha - \sum_{n \neq m} e^{i\mathbf{k}\cdot\mathbf{R}_{nm}}\beta_n(\mathbf{R}_{nm}), \tag{14.3}$$

where

$$\left.\begin{aligned}
\alpha &= -\int |\chi(\mathbf{r} - \mathbf{R})|^2\, U'(\mathbf{r} - \mathbf{R})\, d\mathbf{r}; \\
\beta(\mathbf{R}) &= -\int \chi^*(\mathbf{r} - \mathbf{R})\, U'(\mathbf{r} - \mathbf{R})\,\chi(\mathbf{r} - \mathbf{R})\, d\mathbf{r}; \\
U'(\mathbf{r} - \mathbf{R}) &= V(\mathbf{r}) - U(\mathbf{r} - \mathbf{R}); \quad \mathbf{R} = \mathbf{R} - \mathbf{R}.
\end{aligned}\right\} \tag{14.4}$$

This calculation is valid for the simplest case when a unit cell contains one atom, and one atomic function for each atom is included in the Bloch sum. This is possible only for nondegenerate atomic levels with a sufficiently wide separation between neighboring levels $E_\nu^0 - E_{\nu'}^0$.

In general, the crystal function is a sum of several Bloch sums

$$\psi_k = \frac{1}{\sqrt{G}} \sum_{\mu=1}^{\mu_1} \sum_{\nu=1}^{\nu_\mu} \sum_n A_{\mu\nu k}e^{i\mathbf{k}\cdot\mathbf{R}_n}\chi_{\mu\nu}(\mathbf{r} - \mathbf{R}_{n\mu}), \tag{14.5}$$

where μ is the number of a given atom in a unit cell; μ_1 is the total number of atoms in that cell; ν is the number of atomic functions of a particular atom; ν_μ is the number of the functions of the μ-th atom; $\mathbf{R}_n$ is the radius vector of the central atom and $\mathbf{R}_{n\mu}$ is the radius vector of any μ-th atom in the cell. The energy $E_\mathbf{k}$ is determined from the secular equation

$$\left| M_{\mu\nu\mu'\nu'} - E_\mathbf{k}\delta_{\mu\mu'}\delta_{\nu\nu'} \right| = 0, \tag{14.6}$$

where the matrix elements are

$$M_{\mu\nu\mu'\nu'} = \frac{1}{G} \sum_{n,\,m} e^{i\mathbf{k}\cdot(\mathbf{R}_m - \mathbf{R}_n)} \int \chi^*_{\mu\nu}(\mathbf{r} - \mathbf{R}_{n\mu}) \times \tag{14.7}$$

$$\times U'(\mathbf{r} - \mathbf{R}_{m\mu'})\chi_{\mu'\nu'}(\mathbf{r} - \mathbf{R}_{m\mu'})\,d\mathbf{r} + E^0_{\mu\nu}\,\delta_{\mu\mu'}\delta_{\nu\nu'}$$

The method of tightly bound electrons is most convenient in calculations of the inner shell function of atoms in a crystal and for ionic crystals in which ions have filled electron shells.

In the cell method, the total space is divided into polyhedral cells, surrounding each atom and formed by planes passing through the middles of the segments joining the central atom and its neighbors. Inside each cell, the potential is determined mainly by the central atom and is assumed to be approximately spherically symmetrical, V(r). In this case, the crystal wave function inside each cell may be expanded in terms of spherical harmonics near the center of the cell

$$\psi_k(\mathbf{r}) = \frac{1}{r} \sum_{l,\,m} b_{l,\,m} Y_l^m(\theta,\,\varphi) f_l(r,\,E_k). \tag{14.8}$$

where $Y_l^m(\theta,\varphi)$ is a spherical harmonic, and $f_l(r, E_k)$ is a radial function satisfying the condition

$$\frac{d^2f}{dr^2} - \frac{l(l+1)}{r^2} f + \frac{2m}{\hbar^2}[E_k - V(r)]f = 0. \tag{14.9}$$

The coefficients should be determined from the boundary conditions at the surface of the cell; in the case of multiatomic crystals, these are: the continuity equations for a wave function and for its derivative on transferring from one cell to the next, and Bloch's conditions of Eq. (13.6).

In practice, we must restrict ourselves to a small number of terms in the expansion of (14.8), satisfying the boundary conditions only at selected

spots on the polyhedron surface, lying mainly at the centers of the faces between the nearest neighbors. A more accurate calculation can be carried out
at symmetrical points of the **k** space where some of the harmonics are excluded from the expansion (14.8) on the strength of symmetry conditions.

To use fully the crystal symmetry, the expansion (14.8) is rewritten
in the form:

$$\psi_{ks} = \frac{1}{r} \sum_{l,\,t} B_{lt} K_{slt}(\theta,\ \varphi) f_l(r,\ E_{\mathbf{k}}), \qquad (14.10)$$

where K_{slt} is the lattice harmonic, i.e., the linear combination of spherical
harmonics which transform in accordance with one of the irreducible representations of the group of the wave vector **k.** The lattice harmonics for a
cubic crystal are known as cubic harmonics and are tabulated in [24].

The method of cells is easily extended to complex lattices containing
several atoms in their unit cell [23], and is most convenient for crystals with
a large coordination number, in particular for metals.

The method of orthogonalized plane waves (OPW) is an improvement
on the method of expanding a wave function in a Fourier series of plane
waves. The Fourier series expansion gives, in principle, the solution of
Schrödinger's equation for a crystal, but these series converge very slowly so
that the wave function of a crystal near each atomic nucleus is in the nature
of an atomic function; consequently, to describe the behavior of the wave
function near nuclei, the Fourier series should contain terms with sufficiently
short wavelengths. In order to improve the convergence, we add to each plane
wave terms containing atomic functions, and the coefficients of these terms
are selected so that the total wave is orthogonal to the wave function of the
ionic cores of the atoms. The additional terms describe the behavior of a
wave function near nuclei so that we no longer need to include plane waves
of short wavelengths.

It is more convenient not to add the individual atomic functions of the
cores to these waves but to construct from them, using the tight binding
method, Bloch's functions of the (14.1) type:

$$\varphi_{k\nu}(\mathbf{r}) = \sum_n e^{i\mathbf{k}\cdot\mathbf{R}_n} \chi_\nu(\mathbf{r} - \mathbf{R}_n), \qquad (14.11)$$

where ν is the number of the core function.

The functions $e^{i(\mathbf{k}+\mathbf{K}_i)\cdot\mathbf{r}}$, where $\mathbf{K}_i$ are all possible vectors of the reciprocal lattice translations, represent a complete system of plane waves, suitable for the expansion of the wave function $\psi_{\mathbf{k}}$.

The corresponding orthogonalized plane waves (OPW) have the form:

$$Y_i(\mathbf{k},\ \mathbf{r}) = e^{i(\mathbf{k}+\mathbf{K}_i)\cdot\mathbf{r}} - \sum_\nu \mu_{i\nu}\varphi_{\mathbf{k}+\mathbf{K}_i,\nu}(\mathbf{r}), \qquad (14.12)$$

where

$$\mu_{i\nu} = \int e^{i(\mathbf{k}+\mathbf{K}_i)\cdot\mathbf{r}}\chi_\nu(\mathbf{r})\,d\mathbf{r}. \qquad (14.13)$$

It is easily shown that Y_i is orthogonal to all the ionic functions, i.e.,

$$\int Y_i^*(\mathbf{k},\ \mathbf{r})\,\varphi_{k'\nu}(\mathbf{r})\,d\mathbf{r} = 0 \qquad (14.14)$$

for all $\mathbf{k}'$ and for all levels of the cores, if we assume that the functions χ of various cores are orthogonal to one another.

The wave function of a crystal $\psi_{\mathbf{k}}$ is expanded in a series of OPW

$$\psi_{\mathbf{k}}(\mathbf{r}) = \sum_i \beta(\mathbf{K}_i)\,Y_i(\mathbf{k},\ \mathbf{r}). \qquad (14.15)$$

In this series, we make the approximation of retaining only a finite number of terms; we consider then the coefficients of the expansion $\beta(\mathbf{K}_i)$ as variational parameters and determine them from the condition for a minimum of the integral

$$\int \psi_{\mathbf{k}}^*(\hat{H}-E)\,\psi_{\mathbf{k}}\,d\mathbf{r}; \qquad \hat{H} = -\frac{\hbar^2}{2m}\nabla^2 + V(\mathbf{r}), \qquad (14.16)$$

which, after substitution of (14.15), gives a system of linear equations

$$\sum_i \beta_i \int Y_j^*(\hat{H}-E)\,Y_i\,d\mathbf{r} = 0, \qquad (14.17)$$

and the corresponding secular equation

$$\left|\int Y_j^*(H-E)\,Y_i\,d\mathbf{r}\right| = 0 \qquad (14.18)$$

determines E as a function of $\mathbf{k}$. If we use Eq. (14.12) for Y_i, each element in the determinant becomes a complicated expression so that the OPW calculations are very labor-consuming.

To obtain satisfactory results it is necessary to take a large number of terms in the expansion (14.15). At the symmetrical points of the $\mathbf{k}$ space, the symmetry conditions give relationships between groups of coefficients in (14.15) so that the number of independent β_i and the order of the determinant (14.18) decrease. Complicated calculations using the OPW method are carried out using electronic computers. In practice, the calculations are carried out only for the symmetrical points of $\mathbf{k}$, because for arbitrary points the value of the energy can be obtained by means of tight binding formulas, regarded as interpolation expressions. This means that the integrals in these expressions are found for the symmetrical points determined by the OPW method.

The method of equivalent orbitals is similar to the method of tightly bound electrons, but Bloch's sums are made up not of atomic functions but of "equivalent orbitals," which are wave functions localized near a group of atoms, usually near a covalent bond joining two neighboring atoms. The orbitals are called equivalent because they differ only by their orientation and position in space and are transformed into self-coincidence by symmetry elements of the crystal (translation and rotation).

The method of equivalent orbitals is applicable to crystals, mainly semiconductors, with predominantly covalent binding. Here, the equivalent orbitals are associated with the covalent bonds. Hall [25] used this method to deal with the diamond lattice, Samoilovich and Tovstyuk [26] investigated germanium and silicon, and Gubanov and Nran'yan [27] dealt with semiconductors of the $A^{III}B^{V}$ type. The advantages of the equivalent orbital method compared with the method of tightly bound electrons are as follows:

1) in principle, we may assume that the functions for valence bonds are known more exactly than linear combinations of atomic functions, i.e., the strongest interaction between neighboring atoms is included in the zeroth approximation;

2) the secular equation (14.6) is of lower order; for example, it is quartic for diamond-type lattices and not of the eighth order, because a system of equations splits into a system of binding functions and antibinding functions of valence bonds;

3) the interaction integrals may be obtained from the experimental data as in [25].

In calculating the filled band of a semiconductor, when, as shown in § 11, the Bloch and the Heitler—London methods become identical, the equivalent orbital method is deduced directly from the Heitler—London method [25] or from the elementary excitation method [26].

In conclusion, it should be noted that in all these calculation methods the selection of the potential $V(\mathbf{r})$ is very important because this selection governs to a large degree the accuracy of the results obtained.

§15. Calculation of the Mean Free Path of Electrons

In an ideal crystal lattice, an electron moves without difficulty and its mean free path is infinite. The electron scattering and finite mean free path are due to departures from the periodicity in the atomic distribution: thermal vibrations, defects, and impurities.

At room temperature, the scattering on thermal vibrations is of prime importance, and we shall consider it in some detail. The thermal vibrations of a lattice represent the superposition of elastic waves, each of which is a quantum oscillator. If we assume that the interaction of the electrons with the elastic vibrations (phonons) is weak, then in the zeroth-order approximation the eigenfunction of the system in the Heisenberg representation is equal to a Bloch wave function of an electron $\psi_{\mathbf{k}}(\mathbf{r})$ multiplied by the product of the wave functions of the lattice oscillators $\psi_{N_{\mathbf{q}j}}(a_{\mathbf{q}j})$

$$\Psi(\mathbf{k}, N) = \psi_{\mathbf{k}}(r) \prod_{\mathbf{q}, j} \psi_{N_{\mathbf{q}j}}(a_{\mathbf{q}j}), \qquad (15.1)$$

and the displacement of a lattice point $\mathbf{r}$ by these thermal vibrations is

$$\mathbf{u} = \frac{1}{\sqrt{G}} \sum_{\mathbf{q}} \sum_{j=1}^{3} \mathbf{e}_{\mathbf{q}j} (a_{\mathbf{q}j} e^{i\mathbf{q}\cdot\mathbf{r}} + a_{\mathbf{q}j}^{*} e^{-i\mathbf{q}\cdot\mathbf{r}}); \qquad (15.2)$$

here $\mathbf{e}_{\mathbf{q}j}$ is a unit vector representing the polarization of a phonon having a wave number $\mathbf{q}$; G is the number of atoms in a crystal; and $N_{\mathbf{q}j}$ is the quantum number of a given oscillator.

The wave function of a perturbed system can be expanded as a series of the zeroth-approximation functions in the Heisenberg representation:

$$\Psi(t) = \sum c(\mathbf{k}, N_{\mathbf{q}j}, t) \Psi(\mathbf{k}, N_{\mathbf{q}j}) e^{-\frac{i}{\hbar}\left[E_{\mathbf{k}} + \sum_{\mathbf{q}, j}\left(N_{\mathbf{q}j} + \frac{1}{2}\right)\hbar\omega_{\mathbf{q}j}\right]}; \qquad (15.3)$$

here $E_{\mathbf{k}}$ is the electron energy; $\left(N_{\mathbf{q}j} + \frac{1}{2}\right)\hbar\omega_{\mathbf{q}j}$ is the oscillator energy; and $\omega_{\mathbf{q}j}$ is the eigenfrequency of the oscillator.

The expansion coefficient, according to the quantum transition method, depends on time t and, for small values of t, satisfies the equation:

$$i\hbar\dot{c}(\mathbf{k'}, N'_{\mathbf{q}j}) = \int \psi^* (\mathbf{k'}, N'_{\mathbf{q}j}) U \psi (\mathbf{k}, N_{\mathbf{q}j}) d\tau \times$$
$$\times e^{\frac{i}{\hbar} \left[E_{\mathbf{k'}} - E_{\mathbf{k}} + \sum_{\mathbf{q}, j} \left(N'_{\mathbf{q}j} - N_{\mathbf{q}j} \right) \hbar \omega_{\mathbf{q}j} \right] t}, \tag{15.4}$$

where the integration with respect to $d\tau$ includes integration both with respect to the radius vector $\mathbf{r}$ of the electrons and with respect to the oscillator coordinate $a_{\mathbf{q}j}$; U is the perturbation potential calculated precisely by Bardeen [28] using the self-consistent field method. His result has been found to be close to the simple expression obtained on the assumption of deformable ions made by Bethe and Sommerfeld [29], i.e.,

$$U = -\mathbf{u}\nabla V, \tag{15.5}$$

where V is a periodic self-consistent potential of an electron. Substituting Eqs. (15.1), (15.2), and (15.5) into the integral of Eq. (15.4), and bearing in mind the orthonormalized nature of the functions $\psi_{N_{\mathbf{q}j}}$, we find that this integral becomes:

$$J = -\frac{1}{\sqrt{G}} \sum_{\mathbf{q}, j} \mathbf{e}_{\mathbf{q}j} [\mathbf{K}^+ (a_{\mathbf{q}j})_{N'N} + \mathbf{K}^- (a_{\mathbf{q}j})_{N'N}], \tag{15.6}$$

where

$$(a_{\mathbf{q}j})_{N'N} = \int \psi_{N'_{\mathbf{q}j}} (a_{\mathbf{q}j}) a_{\mathbf{q}j} \psi_{N_{\mathbf{q}j}} (a_{\mathbf{q}j}) d a_{\mathbf{q}j}, \tag{15.7}$$

$$\mathbf{K}^{\pm} = \int \psi^*_{\mathbf{k'}} (\mathbf{r}) \nabla V e^{\pm i\mathbf{q} \cdot \mathbf{r}} \psi_{\mathbf{k'}} (\mathbf{r}) d\mathbf{r}. \tag{15.8}$$

However, $(a_{\mathbf{q}j})_{N'N}$ and $(a^*_{\mathbf{q}j})_{NN'}$ vanish unless $N'_{\mathbf{q}j} = N_{\mathbf{q}j} \pm 1$ and the corresponding matrix elements are

$$(a_{\mathbf{q}j})_{N-1, N} = \sqrt{\frac{\hbar}{2M\omega_{\mathbf{q}j}} N_{\mathbf{q}j}} ; \left. \right\}$$
$$(a^*_{\mathbf{q}j})_{N+1, N} = \sqrt{\frac{\hbar}{2M\omega_{\mathbf{q}j}} (N_{\mathbf{q}j} + 1)} ; \left. \right\} \tag{15.9}$$

M is the atomic mass.

After replacing $\psi_{\mathbf{k}'}$ and $\psi_{\mathbf{k}}$ with Bloch's functions of Eq. (13.9), the integrals involving electron coordinates may be represented in the form of a sum of integrals over all unit cells:

$$\mathbf{K}^{\pm} = \frac{1}{G} \sum_n e^{i(\mathbf{k}\pm\mathbf{q}-\mathbf{k}')\cdot\mathbf{R}_n} \int_1 e^{i(\mathbf{k}\pm\mathbf{q}-\mathbf{k}')(\mathbf{r}-\mathbf{R}_n)} \nabla V u_k u_{k'}^* d\mathbf{r}; \quad (15.10)$$

$1/G$ appears in the above equation because u_k is normalized for a unit cell, and $\psi_{\mathbf{k}}$ for the whole crystal; $\mathbf{R}_n$ is the radius vector of a site in the n-th unit cell and the subscript "1" of the integral denotes that it is taken over one unit cell. This integral is obviously independent of the serial number of the cell and it can be placed outside the sum. The remaining sums for a perfect lattice vanish only if the coefficient of $\mathbf{R}_n$ in the power of the exponential function is zero or equal to the reciprocal-lattice vector $\mathbf{K}$. Hence, we obtain the law of conservation of momentum for the scattering of electrons on phonons

$$\mathbf{k}' = \mathbf{k} \pm \mathbf{q} + \mathbf{K}; \quad (15.11)$$

here the sign "+" represents the absorption of a phonon and "−" represents the emission. The processes with $\mathbf{K} \neq 0$ are known as umklapp processes. We shall only consider the case $\mathbf{K} = 0$. When the conditions of Eq. (15.11) are satisfied, we have

$$\mathbf{K}^{\pm} = \int_1 u_{\mathbf{k}}(\mathbf{r}) \nabla V u_{\mathbf{k}'}^*(\mathbf{r}) d\mathbf{r}. \quad (15.12)$$

If we assume approximately that $E_{\mathbf{k}}$ depends only on $|\mathbf{k}|$, and that $u_{\mathbf{k}}$ and $u_{\mathbf{k}'}$ are spherically symmetrical and equal, and if we use Schrödinger's equations for $u_{\mathbf{k}}$ and $u_{\mathbf{k}'}^*$, and neglect the terms with $E_{\mathbf{k}} - E_{\mathbf{k}'}$ and $\frac{\hbar^2}{2m}(k^2 - k'^2)$ compared with $E_{\mathbf{k}}$, we find (for details see [29]) that

$$\mathbf{e}_{qj} \cdot \mathbf{K}^{\pm} = \pm \frac{i\hbar^2}{m} \mathbf{e}_{qj} (\mathbf{k}' - \mathbf{k}) \int_1 \left| \frac{\partial u_k}{\partial s} \right|^2 d\mathbf{r}, \quad (15.13)$$

where the derivative with respect to s is the derivative with respect to the direction of polarization of $\mathbf{e}_{qj}$. For transverse waves $j = 2, 3$ and Eq. (15.13) is equal to zero, but for longitudinal waves

$$\mathbf{e}_{qj} \cdot \mathbf{K}^{\pm} = \pm \frac{2i}{3} qC, \quad (15.14)$$

where

$$C = \frac{3}{2}\frac{\hbar^2}{2m}\int_1 \left|\frac{\partial u}{\partial s}\right|^2 d\mathbf{r} = \frac{\hbar^2}{2m}\int_1 |\nabla u|^2 d\mathbf{r}; \qquad (15.15)$$

C is the average kinetic energy of a conduction electron.

Substituting Eqs. (15.6), (15.9), and (15.14) into Eq. (15.4) and integrating with respect to time, we obtain

$$|c(\mathbf{k} \pm \mathbf{q},\ N_q \pm 1,\ t)|^2 = \frac{2C^2 q^2}{9GM\hbar\omega_q}\left(N_q + \begin{Bmatrix}1\\0\end{Bmatrix}\right) \times \qquad (15.16)$$

$$\times \,\Omega\,(E_{\mathbf{k} \pm \mathbf{q}} - E_{\mathbf{k}} \pm \hbar\omega_q),$$

where

$$\Omega\,(x) = 2\frac{1 - \cos\dfrac{xt}{\hbar}}{\left(\dfrac{x}{\hbar}\right)^2}. \qquad (15.17)$$

For sufficiently large values of t, $\Omega(x)$ is similar to the function $\delta(x)$, i.e., it is small when $x \neq 0$ and large when $x = 0$, i.e., when

$$E_{\mathbf{k} \pm \mathbf{q}} = E_{\mathbf{k}} \pm \hbar\omega_q. \qquad (15.18)$$

The preceding equation is the law of conservation of energy for the emission and absorption of phonons by electrons. The scattering of an electron is probable only if Eq. (15.18) is satisfied.

Equation (15.16) gives the probability that an electron will go over to the state $\mathbf{k} \pm \mathbf{q}$, and a lattice wave will absorb an energy quantum if at $t = 0$ the electron was definitely in the state $\mathbf{k}$ and N_q quanta were excited.

In order to determine the change in the distribution function of electrons $f(\mathbf{k})$ under the action of collisions, it is necessary to average.out Eq. (15.16) over all possible initial field states. Usually it is assumed that a lattice is in the state of thermal equilibrium so that averaging gives simply the statistical mean N_q. Next, Eq. (15.16) must be multiplied by the probability of the initial electron state $f(\mathbf{k})$ and the probability of a vacant final state, i.e., by $1 - f(\mathbf{k} \pm \mathbf{q})$.

Consequently, differentiating with respect to t, we obtain, after making some transformations,

$$\left(\frac{\partial f}{\partial t}\right)_{\text{collision}} = \frac{2C^2}{9M\hbar G} \cdot \frac{\partial}{\partial t} \sum_{q} \frac{q^2}{\omega_q} \{\Omega\left(E_{k+q} - E_k - \hbar\omega_q\right) \times$$

$$\times \left[(1 - f(k)f(k+q)(N_q + 1) - f(k)(1 - f(k+q))N_q\right] +$$

$$+ \Omega\left(E_{k+q} - E_k + \hbar\omega_k\right) \times$$

$$\times \left[(1 - f(k))f(k+q)N_{-q} - f(k)(1 - f(k+q))(N_{-q} + 1)\right]\}.$$

$$(15.19)$$

The first term in each square bracket gives the number of electrons which are transferred by collisions from other states to k, and the second terms gives the number of electrons ejected from the state k. It is easily seen that if the phonons obey the Bose—Einstein distribution, i.e.,

$$N_q = \frac{1}{e^{\frac{\hbar\omega_q}{k_B T}} - 1}, \qquad (15.20)$$

and if the electrons obey the Fermi-Dirac distribution, Eq. (15.19) vanishes. In the presence of an external field, the electron distribution function departs from its equilibrium form. If the field is not too strong, the departure is not too great and the distribution function may be written in the form:

$$f = f_0 - k_x \frac{\partial f_0}{\partial E} \chi(E); \qquad f_0 = \frac{1}{e^{\frac{E-\mu}{k_B T}} + 1}. \qquad (15.21)$$

In the steady state

$$\frac{\partial f}{\partial t} = \left(\frac{\partial f}{\partial t}\right)_{\text{collision}} + \left(\frac{\partial f}{\partial t}\right)_{\text{field}} = 0, \qquad (15.22)$$

and, if the field $\mathscr{E}$ is directed along the x axis, then

$$\left(\frac{\partial f}{\partial t}\right)_{\text{field}} = -e\mathscr{E} \frac{df_0}{dt} v_x. \qquad (15.23)$$

We shall introduce in the q space a spherical system of coordinates q, ϑ, φ with its axis along k and replace the summation with respect to q by integration, using the formula

$$\sum F(q) = \frac{\Omega_0 G}{(2\pi)^3} \int F(q) q^2 dq \sin \vartheta d\vartheta d\varphi, \qquad (15.24)$$

where Ω_0 is the volume of a unit cell.

We shall assume that q is small and $E_k = E(|\mathbf{k}|)$, and expand $E_{\mathbf{k+q}}$ in series of powers of $(k \pm q)^2 - k^2 = q^2 \pm 2qk \cos \vartheta$. Because of the δ nature of the function Ω, when integrating with respect to ϑ all the remaining factors can be placed outside the integral and the integration gives

$$\int_{-1}^{1} \Omega \left(E_{\mathbf{k\pm q}} - E_{\mathbf{k}} \pm \hbar\omega_q \right) d \cos \vartheta = \frac{2\pi\hbar t}{\left(\dfrac{dE}{dk}\right) q}. \tag{15.25}$$

Substituting Eqs. (15.20), (15.21), (15.24), and (15.25) into Eq. (15.19), and then Eqs. (15.19) and (15.23) into Eq. (15.22), integrating with respect to φ, and using the expansion of $E_{\mathbf{k+q}}$ and Eq. (15.18), we obtain, after making some transformations [29], Bloch's integral equation for $\chi(E)$:

$$\int_0^{q_0} q^2 dq N(q) \left\{ \left[\left(1 + \frac{\hbar\omega}{k\dfrac{dE}{dk}} - \frac{q^2}{2k^2} \right) \chi(E + \hbar\omega) - \chi(E) \right] \times \right.$$

$$\times \frac{f_0(E + \hbar\omega)}{f_0(E)} e^{\frac{\hbar\omega}{k_БT}} + \left[\left(1 - \frac{\hbar\omega}{k\dfrac{dE}{dk}} - \frac{q^2}{2k^2} \right) \chi(E - \hbar\omega) - \chi(E) \right] \times$$

$$\left. \times \frac{f_0(E - \hbar\omega)}{f_0(E)} \right\} = -\frac{9\pi Mwe\mathcal{E}}{\hbar C^2 \Omega_0 k} \left(\frac{dE}{dk} \right)^2; \tag{15.26}$$

here q_0 is the limiting value of q for phonons; and w is the velocity of sound. It is assumed that $\omega = wq$.

Equation (15.26) is solved for several special cases and the quantiy $\chi(E)$ is calculated. According to the free electron theory it is related in the following way to the mean free path of electrons:

$$l = \frac{k}{e\mathcal{E}} \chi(E). \tag{15.27}$$

Details of these calculations are given in Sects. 37 and 45 of Bethe and Sommerfeld's book [29]; we shall present here only the final results.

For metals at $T \gg \theta$, where

$$\theta = \frac{\hbar\omega q_0}{k_B} \tag{15.28}$$

is the Debye temperature,

$$l = \frac{1}{\pi^3} \cdot \frac{M}{\hbar^3} \cdot \frac{k^2}{C^2} \left(\frac{dE}{dk}\right)^2 \mathcal{Q}_0 k_B \theta \frac{\theta}{T} \; ; \tag{15.29}$$

here k_B is Boltzmann's constant.

At $T \ll \theta$

$$l = \frac{1}{4\pi^3 \cdot 124.4} \cdot \frac{M}{\hbar^2} \left(\frac{k}{C} \cdot \frac{dE}{dk}\right)^2 \mathcal{Q}_0 k_B \theta \left(\frac{\theta}{T}\right)^5 . \tag{15.30}$$

For covalent semiconductors at practically all temperatures

$$l = \frac{9 \left(\frac{4\pi}{3}\right)^{2/3} M a k_B \theta}{16\pi \hbar^2 C^2 (ak)^2} \left(\frac{dE}{dk}\right)^2 ; \tag{15.31}$$

here a is the lattice constant, or more precisely $a = \sqrt[3]{\mathcal{Q}_0}$.

In ionic semiconductors, electrons are scattered mainly on the optical modes of lattice vibrations. The perturbation energy, according to Fröhlich [30], has the form:

$$U = -\frac{4\pi i e^2}{2a^3 q \sqrt{2G}} \sum_q (a_q e^{i q \cdot r} - a_q^* e^{-i q \cdot r}), \tag{15.32}$$

and irrespective of the condition (15.11) only the longitudinal vibrations inter-act with electrons. The frequency of the optical-mode vibrations depends weakly on q and is assumed to be equal to a constant value ω_0 so that the appropriate factors may be taken outside the integral sign when integrating with respect to q.

In other respects, the calculations are the same as before [31], and they give, at $T \gg \theta$, where

$$\theta = \frac{\hbar \omega_0}{k_B} , \tag{15.33}$$

$$l = \frac{a}{2\pi} \cdot \frac{M}{m^*} \left(\frac{\hbar \omega_0}{\frac{e^2}{a}}\right)^2 \frac{E}{k_B T} , \tag{15.34}$$

and at $T \ll \theta$

$$l = \frac{a}{2\pi} \cdot \frac{M}{m^*} \cdot \frac{\hbar\omega_0}{\frac{e^2}{a}} e^{\frac{\hbar\omega_0}{k_B T}} \sqrt{\frac{E}{\hbar\omega_0}}. \tag{15.35}$$

The calculation of the electron scattering on stationary departures from periodicity (defects, inhomogeneous regions in disordered alloys, etc.) is simpler.

The perturbation potential is simply a deviation from the periodic potential. In the case of disordered alloys, the perturbation potential for an atom is

$$U = U_n(\mathbf{r} - \mathbf{R}_n) - \sum_s p_s U_s - (\mathbf{r} - \mathbf{R}_n) = V_n - V_0, \tag{15.36}$$

where p_s is the fraction of atoms of type s; U_s is the potential of atoms of type s; n is the serial number of an atom; and V_0 is the atomic potential averaged out over all possible types of atom.

In all these formulas, $\hbar\omega$, $\mathbf{q}$, and N are eliminated and the summation may be carried out with respect to $\mathbf{q}$, and not $\mathbf{k}'$. Instead of Eq. (15.19) we now have

$$\left(\frac{\partial f}{\partial t}\right)_{\text{alloy}} = \frac{\partial}{\partial t} \sum_{\mathbf{k}'} \frac{|U_{\mathbf{k}'\mathbf{k}}|^2}{\hbar^2} \Omega (E_{\mathbf{k}'} - E_{\mathbf{k}}) \times \tag{15.37}$$
$$\times \{f(\mathbf{k}')[1 - f(\mathbf{k})] - f(\mathbf{k})[1 - f(\mathbf{k}')]\};$$

here the matrix element $U_{\mathbf{k}'\mathbf{k}}$ can be represented in the form

$$U_{\mathbf{k}'\mathbf{k}} = \int \psi_{\mathbf{k}'}^* U \psi_{\mathbf{k}} d\mathbf{r} = \frac{1}{G} \sum_n e^{i(\mathbf{k}-\mathbf{k}') \cdot \mathbf{R}_n} (U_n^{\mathbf{k}\mathbf{k}'} - U_0^{\mathbf{k}\mathbf{k}'}), \tag{15.38}$$

where

$$U_n^{\mathbf{k}\mathbf{k}'} = \int_1 U_n(\mathbf{r}) u_{\mathbf{k}'}^*(\mathbf{r}) u_{\mathbf{k}}(\mathbf{r}) e^{i(\mathbf{k}-\mathbf{k}') \cdot \mathbf{r}} d\mathbf{r}. \tag{15.39}$$

Changing from summation to integration in a spherical system of co-ordinates k', ϑ, and φ with its axis along k, assuming that $U_{\mathbf{k}\mathbf{k}'} = V(\vartheta)$, i.e., that the matrix element depends only on the angle ϑ between $\mathbf{k}$ and $\mathbf{k}'$ and not on $\mathbf{k}$ and $\mathbf{k}'$ themselves, and substituting Eq. (15.21), we integrate with respect to k' and φ and find

$$\left(\frac{\partial f}{\partial t}\right)_{\text{alloy}} = \frac{G\Omega}{2\pi\hbar} \cdot \frac{k^2}{\frac{dE}{dk}} \cdot \frac{df_0}{dE} \chi(E) \times \int_0^\pi |V(\vartheta)|^2 (1 - \cos\vartheta) \sin\vartheta d\vartheta.$$

(15.40)

The substitution of Eq. (15.40) into equations of the (15.22) type and the solution of the resultant equation gives, according to Eq. (15.27), the electron mean free path

$$l = \frac{2\pi}{a^2 k^2} \left(\frac{dE}{dk}\right)^2 \frac{1}{\int_0^\pi \overline{|U_n^{k'k}|^2} (1 - \cos\vartheta) \sin\vartheta d\vartheta} ;$$

(15.41)

the bar denotes averaging over all atoms.

The theory of electron scattering in a crystal has been extended to a more complex dependence $E(\mathbf{k})$ [32] but we have deliberately presented the older theory assuming the isotropy of the dependence $E(\mathbf{k})$ because it is precisely this theory which can be applied to liquids.

§ 16.　Impurity Levels and Local Electron States

If the departure from periodicity in a crystal represents a sufficiently deep additional potential well background, in the periodic potential a local energy level appears; the wave function of this level does not extend over the whole crystal like Bloch's functions but is concentration near the defect and decays away from it, like atomic functions. The local levels are frequently due to impurity atoms and therefore they are known as impurity levels.

The main types of defect which may give rise to local electron states are: a vacancy, a host atom at an interstitial position (an interstitial impurity), and a foreign atom replacing a host atom at a regular site (a substitutional impurity). In the latter case, the impurity atom produces a deep potential well provided the charge of its core is different from the charge of the cores of the host atoms. The role of the impurity atom depends strongly on whether it is located at a regular site or at an interstice.

In the majority of cases, a defect can be considered to be a point charge located in a periodic potential field. A positive charge produces a Coulomb well for electrons, and a negative charge produces a well for holes.

Schrödinger's equation in the presence of a defect is

$$-\frac{\hbar^2}{2m}\,\nabla^2\psi + V(\mathbf{r})\,\psi + U(\mathbf{r})\,\psi = E\psi, \qquad (16.1)$$

where V is a periodic potential, and U is a potential due to the defect, which cannot be regarded as a small perturbation because it is large in a certain part of the crystal space. An exact solution of Eq. (16.1) is difficult even for the simplest potential.

In the case of "shallow" levels, the eigenfunctions of which have a radius large compared with the lattice constant, we may use the effective mass method. The essence of this method is that we omit the term with the periodic potential in Schrödinger's equation and replace the normal electron mass m in the first term with an effective mass m*.

At large distances r from an impurity ion, its potential can be written in the form:

$$U(\mathbf{r}) = -\frac{e^2}{\varepsilon r}\,, \qquad (16.2)$$

where ε is the static permittivity. Equation (16.1) then becomes:

$$-\frac{\hbar^2}{2m^*}\,\nabla^2\psi - \frac{e^2}{\varepsilon r}\,\psi = E\psi. \qquad (16.3)$$

The effective mass method is based on the fact that the motion of an electron in a periodic field is similar to the motion of a free electron with an effective mass different from the free-electron mass. This property is retained in the presence of an additional potential which does not vary greatly within a distance equal to the lattice constant. The mathematical justification of the effective mass method has been given by Kohn [33].

The proof of this method is based on the expansion of the solution of Eq. (16.1) as a series of Bloch's functions, the expansion of the periodic multipliers u_k of Bloch's functions as a Fourier series, and the expansion of the unperturbed state energy E_k as a series of small values of $\mathbf{k}$.

The substitution of these expansions into Eq. (16.1) and its transformation back to the coordinate representation gives Eq. (16.3), which is the equation for a hydrogen-like atom. The energy levels given by this equation are

$$E_n = -\frac{1}{n^2 2\hbar^2 m^*}\left(\frac{e^2}{\varepsilon}\right)^2, \qquad (16.4)$$

and the eigenfunctions ψ_n are hydrogen-like functions. For example, the ground-state function is

$$\psi_i = \frac{1}{(\pi a^*)^{3/2}} e^{-\frac{r}{a^*}}, \qquad (16.5)$$

where the effective Bohr radius a* is

$$a^* = \frac{\hbar^2 \varepsilon}{m^* e^2}. \qquad (16.6)$$

If $\varepsilon \gg 1$, and $m^* \ll m$, then a* is much greater than the lattice constant and this ensures the validity of the effective mass method. These assumptions are valid for donor and acceptor levels in germanium and silicon. The donor level energy is taken from the lower edge of the conduction band, and the acceptor level energy (the hole energy) is taken from the upper edge of a filled band. Equation (16.5) shows that these levels lie very close to the corresponding band edges, in agreement with experiment, i.e., they are shallow levels.

Pekar [34] applied the effective mass method to local electron states of the F-center type, i.e., to electrons localized near vacant anion sites in ionic crystals, and he supplemented this by allowing for the polarization of the lattice by the average electron charge. This enabled him to explain several electrical, photoelectric, and optical effects.

To calculate deeper impurity levels and to deal with complex dependences $E(\mathbf{k})$, a generalized effective mass method has been proposed [35,36] the essence of which is as follows. Let us assume that we know the dependence of the energy on the wave vector $E(\mathbf{k})$ for a lattice without impurities. In this dependence, we replace k with $-i\nabla$ and substitute the resultant differential operator in place of the usual kinetic energy operator in Schrödinger's equation, omitting the term with the periodic potential.

Consequently, we obtain the following equation

$$E(-i\nabla)\psi - \frac{e^2}{\varepsilon r}\psi = E\psi. \qquad (16.7)$$

It is derived from Eq. (16.1) in the same way as Eq. (16.3) except that the expansion of $E_0(\mathbf{k})$ as a series in small values of $\mathbf{k}$ is not carried out. If $E(\mathbf{k}) = \frac{\hbar^2 k^2}{2m^*}$, which applies at an energy band edge in the isotropic approximation, then Eq. (16.7) reduces to Eq. (16.3).

The effective mass method gives satisfactory values of the impurity levels, but the smoothed-out hydrogen-like wave functions, which are solutions of Eq. (16.3) or Eq. (16.7), are insuffficiently accurate for the calculation of the matrix elements for quantum transitions.

Dexter [37], Krumhansl [38], and Deigen [39] have shown that the correct wave functions for impurities are approximately equal to the products of the smoothed-out hydrogen-like functions and Bloch's functions corresponding to the edge of a conduction band (for donors) or a valence band (for acceptors).

To calculate deep local levels, whose wave functions have small radii and are localized mainly in one unit cell, a different method is more convenient: it was proposed by Gourary and Adrian [40] for the F centers in alkali-halide crystals. The wave functions are calculated by a variational method using the Madelung potential produced by all lattice ions regarded as point charges. The cubic symmetry of these crystals is taken into account and the wave functions are made orthogonal to the wave functions of the surrounding ions. Further improvement of this theory, proposed by Aborenkov [41], consisted of allowing for the polarizability of a lattice consisting of point charges and for the displacement of ions nearest to the defect.

Chapter III

STRUCTURE OF AMORPHOUS SUBSTANCES

§17. Phenomenological Similarity of Liquids and Solids

Before applying the solid-state theory presented in Chap. II to liquids, it is essential to be clear about the similarities and differences between crystals and liquids.

For a long time, liquids have been regarded as closer in structure to gases than to solids. This is partly due to the successes of the van der Waals theory, which describes the continuous transition from gas to liquid, and partly due to the fact that the mechanical properties of gases and liquids can be described by the same equations of hydrodynamics. Liquids and gases flow while solids shear.

In fact, the difference between solids and liquids is not as great as it would seem at first sight. Crystals flow at high pressures and high shear stresses. On the other hand, liquids have a shear modulus which is normally masked by the flow properties but it does appear at high-frequency vibrations. Finally, in the case of glasses a continuous transition is observed from the mechanical properties of flowing liquids.

The similarity of liquids and solids is even greater in other properties. For example, the volume does not change more than by 10% on melting, indicating similar particle distribution in solids and liquids. The latent heat of fusion is usually small compared with the latent heat of vaporization; for example, the heats of fusion of sodium, zinc, lead, and mercury are, respectively, 630, 1800, 1170, and 560 cal·(g-atom)$^{-1}$, while the heats of vaporization of the same metals are 23,300, 27,730, 46,000, and 14,200 cal·(g-atom)$^{-1}$, i.e., the heats of vaporization are 30-40 times the heats of fusion. This indicates that the vaporization of a substance is a much more radical change of state than the fusion, and that practically the same binding forces act in liquids and solids.

Table 6, taken from Frenkel's book [1], lists the specific heats of some metals, diatomic gases, and compounds in the liquid and solid state. It

70

TABLE 6. Specific Heats of Some Substances in the Solid and Liquid States (cal·deg^{-1}·mole^{-1})

Parameter	Na	Hg	Pb	Zn	Al	H$_2$	Cl$_2$	Br$_2$	CH$_4$	HCl	NH$_3$	C$_6$H$_6$
$(C_p)_s$	7.6	6.7	7.2	7.2	6.8	11.3	14	14.1	10.6	12.26	12.2	26.6
$(C_p)_l$	8	6.7	7.7	7.9	6.25	13.1	16.2	17.2	13.5	14.73	18.4	30.1

follows from this table that the specific heat is practically unaffected by fusion, i.e., the thermal motion in liquids is of the same nature as in solids.

The latter conclusion seems, at first sight, very surprising. It is known that the specific heat of solids is due to the presence of longitudinal and transverse elastic waves and that transverse waves of audio and ultrasonic frequenices are absent in liquids. This contradiction is resolved purely phenomenologically using the Maxwell model of a liquid, according to which the shear deformation of a liquid consists of an elastic deformation with a shear modulus G and a viscous deformation with a viscosity η. The equation of motion of a Maxwell liquid in the case of small pure-shear vibrations is given in Chap. IV of Frenkel's book [1]:

$$\eta \nabla^2 \mathbf{v} = \rho \left(1 + \tau \frac{\partial}{\partial t} \right) \frac{\partial \mathbf{v}}{\partial t}$$
$$+ \left(1 + \tau \frac{\partial}{\partial t} \right) \nabla p, \qquad (17.1)$$

where $\tau = \eta/G$ is the relaxation time of the medium; $\mathbf{v}$ is the velocity of the liquid; ρ is the density; and p is the pressure.

Equation (17.1) differs from the Navier−Stokes equation for an incompressible liquid by the replacement of η with the operator

$$\frac{\eta}{1 + \tau \frac{\partial}{\partial t}}$$

and by the omission of a term containing the velocity squared (because the vibrations are small).

The solution of Eq. (17.1) may be obtained in the form of a plane wave:

$$\mathbf{v} = \mathbf{v}_0 e^{i(\omega t - kx)} \quad (\mathbf{v} \perp x), \qquad (17.2)$$

where [2]

$$k = k_1 - ik_2 = \omega \sqrt{\frac{\rho}{2\eta}} \left[\sqrt{\sqrt{\tau^2 + \frac{1}{\omega^2}} + \tau} - i \sqrt{\sqrt{\tau^2 + \frac{1}{\omega^2}} - \tau} \right]. \qquad (17.3)$$

The above formula shows that, when $\omega\tau < 1$, the damping of this wave is so strong that in practice transverse waves are absent. Conversely, when $\omega\tau \gg 1$, the damping is weak and transverse waves are propagated.

Experiments on colophony have proved the validity of the Maxwell model, the shear modulus of colophony being $G \approx 10^9$ dyn/cm^2. If we assume the same modulus for normal liquids, for which $\eta \approx 0.01$ poise, we find that $\tau = 10^{-11}$ sec. For ultrasound, ω is 10^5-10^8 sec^{-1}, so that $\omega\tau \ll 1$ and transverse waves are absent. In thermal motion, hypersonic waves are mainly active: their frequencies are 10^{12}-10^{13} sec^{-1} so that for them $\omega\tau \gg 1$, i.e., transverse thermal waves exist in liquids.

Equation (17.1) can be used to investigate continuous transition from the mechanical properties of liquids to those of solids as the frequency is increased.

§18. Molecular Structure of Liquids and Structural Diffusion

In contrast to crystals, there is no long-range order in the atomic distribution in liquids, but many experimental observations indicate the presence of short-range order. The degree of short-range order may be judged from the distribution of atoms around another given atom described by a distribution density function $\rho(r)$, which is equal to the average number of atoms in the unit volume defined by the sphere with a radius r taken from the central atom. The product

$$4\pi r^2 \rho(r)\, dr = g(r)\, dr \qquad (18.1)$$

defines the average number of atoms in a spherical layer from r to r + dr.

Such a description of the atomic distribution is also possible for crystals. In the absence of thermal motion, the centers of the atoms are fixed at definite points and the radial distribution function has the form:

$$g(r) = \sum_s z_s \delta(r - r_s), \qquad (18.2)$$

where $\delta(x)$ is the well-known Dirac delta function; r_1, r_2, etc., are the coordinates of the first, second, etc., nearest neighbors, and z_1, z_2, etc., are the numbers of such neighbors.

Under the action of thermal motion the atomic coordinates vary within certain limits, oscillating about the equilibrium positions, and the discrete distribution is smeared out so that the δ functions in Eq. (18.2) must be replaced with functions having finite maxima, for example, with the Gaussian functions:

$$g(r) = \sum_s \frac{z_s}{\sqrt{2\pi \overline{y_s^2}}} e^{-\frac{(r-r_s)^2}{2\overline{y_s^2}}} ; \qquad (18.3)$$

here $\overline{y_s^2}$ is the mean square of the displacement of an atom from its equilibrium position. If atoms are bound to their equilibrium positions by quasielastic forces of rigidity f, we can easily show, using the Boltzmann distribution, that

$$\overline{y_s^2} = \frac{k_B T}{f} \qquad (18.4)$$

and is independent of s. Consequently, as r increases, the widths of the Gaussian distribution maxima, equal to $\sqrt{\overline{y_s^2}}$, remain constant so that the distances between the maxima decrease (this follows from simple geometrical considerations). Therefore, as r increases, these maxima overlap more and more, forming finally a continuous background which is represented by $\rho = g/4\pi r^2$ and gives the average concentration of atoms in a crystal.

Prins [3] (see also [1]) suggested that a density function be derived in the same way for a liquid but that the Gaussian curves in this instance must have maxima of width $\sqrt{\overline{y_s^2}}$, which is different from that for crystals, and which increases gradually with increasing r_s. This gradual broadening of the maxima is not due to an increase of the amplitude of thermal vibrations about their equilibrium positions, but to an increase of the random scatter of the distribution of the equilibrium positions themselves due to the absence of long-range order in liquids.

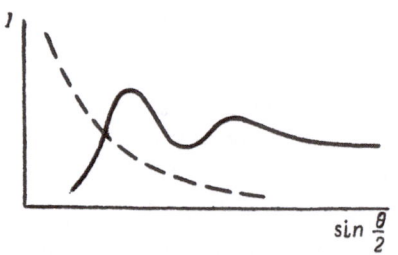

Fig. 10. X-ray intensity distribution as
a function of the scattering angle.

It follows from general statistical considerations that the width of a maximum should increase as the square root of the distance of the appropriate atomic layer from the central atom, i.e.,

$$\overline{y_s^2} = 2Dr_s, \qquad (18.5)$$

where D is a constant having the dimensions of length. Equation (18.5) follows directly from Lyapunov's theorem on the summation of random quantities and it is similar to the expression obtained in the diffusion theory for the mean square of the displacement of a particle in time t_s:

$$\overline{y_s^2} = 2Dt_s, \qquad (18.6)$$

where D is the usual diffusion coefficient. By analogy with the diffusion theory, the coefficient D in Eq. (18.5) is called the structural diffusion coefficient. This coefficient represents the random scatter of the equilibrium positions of atoms when the long-range order is destroyed. The higher the value of D, the lower the degree of short-range order in an amorphous substance. For crystals D = 0, but to make Eq. (18.5) applicable to crystals when D = 0 it must be rewritten in the form:

$$\overline{y_s^2} = \overline{y_0^2} + 2Dr_s, \qquad (18.7)$$

where $\sqrt{\overline{y_0^2}}$ is the broadening of the maxima due to atomic thermal vibrations.

We shall see that the idea of structural diffusion in the relative positions of the atoms in a liquid is confirmed by x-ray diffraction patterns of liquids. The characteristic feature of these patterns, which distinguishes them from those for gases, is the absence of rays scattered at very low angles and the presence of several intensity maxima in the form of rings of increasing diffuseness. Figure 10 shows, by a continuous line, the intensity distribution as a function of the scattering angle θ in an x-ray diffraction pattern of a liquid; the dashed curve represents the pattern for a gas.

The radial distribution function is found [4] from an x-ray diffraction pattern using the formula

$$g(r) = 4\pi r^2 \rho_0(r) + \frac{2r}{\pi} \int_0^\infty i(\varkappa) \varkappa \sin \varkappa r d\varkappa, \qquad (18.8)$$

where

$$\varkappa = \frac{4\pi}{\lambda} \sin \frac{\theta}{2}; \quad i(\varkappa) = \frac{I(\varkappa)}{NF^2} - 1; \qquad (18.9)$$

λ is the wavelength; F is the atomic scattering factor; $I(\varkappa)$ is the intensity of unmodified radiation; and N is the number of atoms.

The $\rho(r)$ curve for a liquid, like the $I(\theta)$ curve, which represents the distribution of intensity in an x-ray diffraction pattern, exhibits maxima of increasing diffuseness, thereby indicating the presence of short-range order in a liquid.

The radial function presents an incomplete description of the atomic distribution because of the averaging over a sphere. To describe the atomic distribution around a central atom more precisely, we introduce a density which is a function not only of r but of all three spherical coordinates, $\rho(r, \vartheta, \varphi)$, with the polar axis directed through the center of one of the nearest neighbors of the central atom.

In a crystal at T = 0

$$\rho(r, \vartheta, \varphi) = \frac{1}{r^2 \sin \vartheta} \sum_s \delta(r - r_s) \sum_{j=1}^{z_s} \delta(\vartheta - \vartheta_{sj}) \delta(\varphi - \varphi_{sj}),$$
$$(18.10)$$

where $\vartheta_{11} = 0$, and the remaining ϑ_{sj} and φ_{sj} depend on the crystal structure; thus, for example, in a simple cubic lattice

$$\vartheta_{12} = \vartheta_{13} = \vartheta_{14} = \vartheta_{15} = \frac{\pi}{2}; \quad \vartheta_6 = \pi; \quad \varphi_{12} = 0; \quad \varphi_{13} = \frac{\pi}{2};$$

$$\varphi_{14} = \pi; \quad \varphi_{15} = \frac{3\pi}{2}.$$

In a crystal at $T > 0$ and in a liquid, we find, in analogy to Eq. (18.3),

$$\rho\,(r,\ \vartheta,\ \varphi) = \frac{1}{r^2 \sin^2 \vartheta} \sum_s \frac{1}{2\pi y_s^2}\, e^{-\frac{(r-r_s)^2}{2\overline{y_s^2}}} \times$$

$$\times \sum_{j=1}^{z_s} B_{sj} e^{-\frac{(\vartheta - \vartheta_{sj})^2 r_s^2}{2\overline{y_s'^2}} - \frac{(\varphi - \varphi_{sj})^2 r_s^2 \sin^2 \vartheta}{2\overline{y_s''^2}}} ; \qquad (18.11)$$

here $\overline{y_s'^2}$ and $\overline{y_s''^2}$ are the averages of the squares of the atomic displace-ments along the axes ϑ and φ. B_{sj} are defined by the condition of normaliza-tion of the Gaussian functions, which is different from the usual condition be-cause ϑ and φ vary within finite limits. This condition is

$$\int_0^{2\pi}\int_0^{\pi} e^{-\frac{(\vartheta - \vartheta_{sj})^2 r_s^2}{2\overline{y_s'^2}} - \frac{(\varphi - \varphi_{sj})^2 r_s^2 \sin^2 \vartheta}{2\overline{y_s''^2}}} \sin \vartheta\, d\vartheta\, d\varphi = \frac{1}{B_{sj}}, \qquad (18.12)$$

and $\overline{y_s'^2}$ and $\overline{y_s''^2}$ are defined by functions similar to Eq. (18.7), i.e.,

$$\overline{y_s'^2} = \overline{y_0'^2} + 2D'r_s, \quad \overline{y_s''^2} = \overline{y_0''^2} + 2D''r_s. \qquad (18.13)$$

It is easily seen using Eq. (18.12) that the functions (18.10) and (18.2) or (18.11) and (18.3) are related by the expression

$$g\,(r) = \int_0^{2\pi}\int_0^{\pi} \rho\,(r,\ \vartheta,\ \varphi)\, r^2 \sin \vartheta\, d\vartheta\, d\varphi, \qquad (18.14)$$

which should be valid for any $\rho(r,\ \vartheta,\ \varphi)$ and $g(r)$.

§19. Structure of Liquid Metals

A systematic presentation of the x-ray diffraction data on various liquids is given in Danilov's book [5]. Liquid metals have been investigated
more than other liquids. A very full review of the x-ray and neutron diffraction data on liquid metals has been given by Radchenko [6], who listed a large

number of x-ray and neutron scattering curves as a function of $\dfrac{\sin \frac{\theta}{2}}{\lambda}$,

and atomic radial distribution curves g(r) or ρ(r) for all the investigated metals. All these curves have several clear maxima, which vary from metal to
metal and depend particularly on the closeness of the crystallographic packing.

Mercury has been investigated for longer and more fully than any other
metal. Debye and Menke [7] were the first to calculate, from the experimental x-ray diffraction curves, the radial distribution function of mercury
atoms; it was found to be very similar to the radial distribution function for
"smeared-out" hexagonal close packing. Danilov and Neimark [8] investigated mercury by x-ray diffraction at low temperatures and found that the
structure of liquid mercury differs from close-packed, but near the crystallization point approaches its solid-state structure, i.e., rhombohedral. Figure 11
shows the radial distribution curves of mercury atoms at various temperatures,
taken from the work of Campbell and Hildebrand [9]. After comparing critically the results of many workers, we may conclude that a little above the
melting point the coordination number of liquid mercury is the same as that
of solid mercury, i.e., six, and that with rise of temperature it increases to
8-10, reaching 12 at high temperatures.

Other metals can be divided into three groups.

1. Metals with close packing. Figure 12 gives, by way of example, the
radial distribution curve for liquid gold. The curves obtained for liquid lead,
aluminum, and zinc are very similar. Gold is an excellent example of the
similarity of the structures in the liquid and solid states: the principal maximum of the intensity curve coincides with the first line of the x-ray diffraction pattern of gold powder; the other maxima coincide with groups of lines;
the packing remains close, and the radius of the first coordination sphere in
liquid gold is equal to the lattice spacing in gold crystals.

Although there are disagreements between the various experimental data
for lead, the majority of investigators are of the opinion that the atomic packing remains close after melting, retaining a coordination number of about 12.
Very nearly the closest possible packing of atoms is also exhibited by liquid
aluminum and zinc.

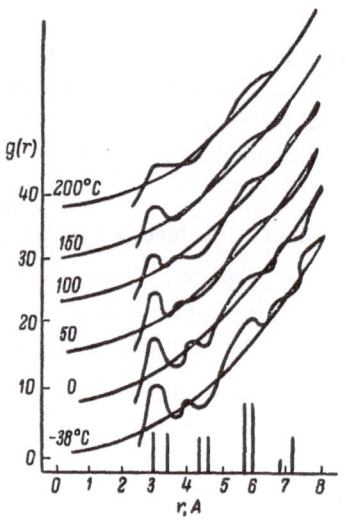

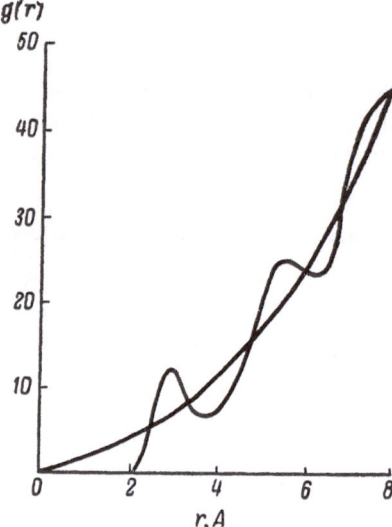

Fig. 11. Radial distribution curves for atoms in liquid mercury at various temperatures. Here, and in Figs. 12-15, the smooth curves represent uniform distribution.

Fig. 12. Radial distribution curve of atoms in liquid gold.

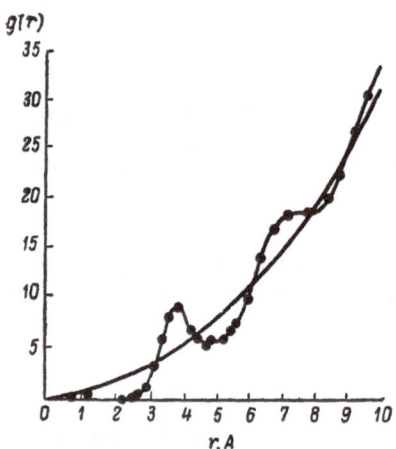

Fig. 13. Radial distribution curve of atoms in liquid sodium.

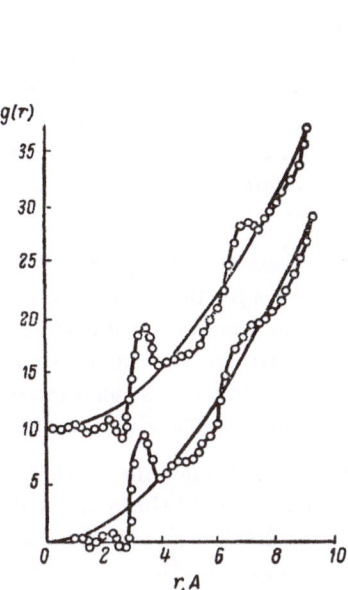

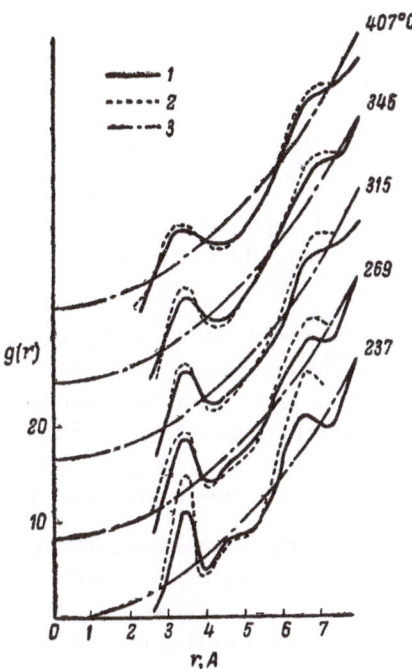

Fig. 14. Radial distribution
curves of atoms in liquid
bismuth at two temperatures.

Fig. 15. Radial distribution curves
of atoms in liquid tin at various
temperatures: 1) experimental
curves; 2) theoretical curves; 3)
uniform distribution curves.

Studies of thallium, indium, and cadmium apparently indicate a change
of the short-range order on melting: the coordination number decreases from
12 to approximately 8. However, these results are not reliable [6].

2. Alkali metals. Liquid sodium has been investigated more than other
alkali metals; its radial distribution curve at 100°C is shown in Fig. 13. On
increase of temperature, the maxima of this curve retain their positions but
they broaden. On melting, the coordination number remains equal to 8, even
after heating to a temperature 300°C above the melting point.

Much less is known about the other alkali metals (lithium, rubidium,
and cesium).

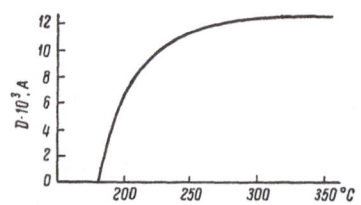

Fig. 16. Temperature dependence of the structural diffusion coefficient D.

3. Metals with loose packing of their atoms in the solid state. In the majority of these metals the structure changes on melting and the coordination number increases. For example, solid gallium has one atom at a distance r = 2.48 A and six atoms at a distance r = 2.71-2.77 A, while liquid gallium has 11 atoms at a distance r = 2.77 A.

The coordination number of bismuth increases on melting from 6 to 7-7.5. Figure 14 shows the radial distribution curves for liquid bismuth at two temperatures; they were obtained by neutron diffraction [10]. Solid and liquid antimony also show a difference in their atomic distributions. Liquid tin [11] exhibits an atomic packing similar to that in tin crystals but on increase of temperature the atomic distribution in the liquid approaches close packing.

From the theoretical point of view, the work of Glauberman and Tsvetkov [12, 13] is of greatest interest: they investigated liquid tin by x-ray diffraction at various temperatures and plotted the g(r) curves; some of these curves are shown in Fig. 15. They found that the experimental curves fitted the theoretical ones for different hexagonal close packings, calculated using

Eqs. (18.3) and (18.7), where $\overline{y_0^2}$ is given by Eq. (18.4) and $f = 0.7 \cdot 10^4$ dyn/cm for tin. The value of D at each temperature was selected to ensure the best agreement between the experimental and theoretical curves. Figure 16 shows the temperature dependence of the structural diffusion coefficient deduced from these data; this coefficient remains almost constant over a wide range of temperatures (Fig. 2 in [12] shows not D but D/π in A).

The work of Glauberman and Tsvetkov is a direct experimental proof of the theory of structural diffusion and it shows that the degree of departure from the short-range order, represented by the structural diffusion coefficient D, is small even when liquid metals are heated to temperatures well above the melting point. It is worth noting that the g(r) curves for other metals have no less sharp maxima, in other words, the conclusions of Glauberman and Tsvetkov [12] apply to all liquid metals.

§20. Structure of Water and of Organic Liquids

The x-ray diffraction patterns of water, like those of metals, exhibit several maxima. The ratios of the amplitudes of these maxima and their different behavior on heating indicate the presence in water of periodic variations due to the ordered distributions of the molecules in small regions.

Bernal and Fowler [14] deduced from x-ray diffraction studies that water has a pseudocrystalline structure and this explains some of its anomalous properties. Water consists of H_2O molecules which are practically identical with those in water vapor. There are three main types of distribution of H_2O molecules: water I, which is like tridymite ice, present in small amounts below 4°C; water II, which is of quartz type and is the predominant modification at normal temperatures; water III, which is an ideal liquid with close packing and predominates at high temperatures. As the temperature is varied, these modifications transform smoothly from one to the other so that water always remains homogeneous without the mixing of volumes having different structures, and only the average mutual positions approach water I, II, or III.

The anomalous compressibility of water down to 4°C is related to the destruction of the loose structure of tridymite and the transition to the closer quartz-type packing. Investigations of the Raman spectra and the dielectric properties have confirmed Bernal and Fowler's ideas presented in [14].

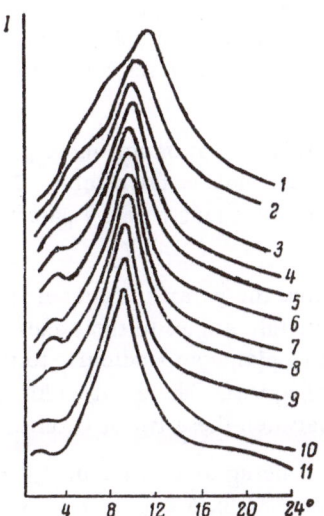

Fig. 17. Intensities of x rays scattered by normal alcohols, as a function of the scattering angle.

To understand the results of x-ray diffraction studies of organic liquids, we must refer to the work of Stewart [15], who showed that the diffraction pattern for organic liquids can be explained neither by a regular distribution of atoms in the molecules nor by an average distance between the molecules: rather, it is due to a certain order in the distribution of the molecules, which Stewart called the cybotactic state of a liquid. Figure 17 shows the scattering curves for normal alcohols (I as a function of the angle). The numbers by the curves denote the number of carbon atoms in an alcohol molecule: from 1 for CH_3OH to 11 for $C_{11}H_{23}OH$. The origin of the ordinate is shifted by one unit for each consecutive curve.

All curves have two maxima representing the molecular distances d_1 and d_2 ($d_1 > d_2$). The principal (second) maximum is due to the transverse dimensions of the molecules $d_2 = 4.6$ A. The position of the first (weaker) maximum depends on the number of carbon atoms in the chain; d_1 rises with the number of atoms but it is numerically greater than the length of the molecule calculated from the density, molecular weight, and transverse dimensions. This means that the first maximum is not due to the effect of the average distance. Hence, we may conclude that both maxima are due to an oriented distribution of the molecules in space with the quantities d_1 and d_2 repeated several times, i.e., regions of oriented molecular distribution exist in these liquids.

These regions are not crystallites because special experiments have shown that the x-ray scattering curves are different for liquids and solids. The first maximum of the solid phase is shifted toward smaller angles compared with the first maximum of the liquid phase; the second maximum is shifted toward larger angles.

From these considerations, Stewart deduced that organic liquids are in the cybotactic state. The characteristic feature of this state is the presence of groups of mutually oriented molecules which have not lost their mobility.

X-ray diffraction patterns of fatty acids, like those of alcohols, have two maxima with the principal maximum again due to the transverse dimensions of the molecules.

Liquid paraffins exhibit only one broad intensity maximum whose position is identical with the principal maximum of alcohols or fatty acids. The absence of the first maximum may be due to the great length of the paraffin molecules.

Stewart has suggested that in ordered regions the paraffin chains lie parallel to one another forming a square network at right angles to their axes. On the other hand, Warren has shown [16] that in a plane perpendicular to the molecular axes the structure is close-packed hexagonal. The great width of the maximum prevents us from distinguishing between these two structures.

Liquid benzene has one maximum, corresponding to a period $d = 4.7$ A, while solid benzene has two very intense lines at $d = 4.82$ A and $d = 4.47$ A, which are close to $d = 4.7$ A. From these data, we may conclude that small regions of liquid benzene retain the same mutual orientations as in a crystal. However, because of the small dimensions of these ordered regions, the x-ray diffraction lines broaden and merge into one maximum.

§21. Structure of Glass

A glass is a supercooled liquid. Therefore, although the mechanical properties of a glass are typical of solids, its molecular structure is the same as that of liquids. In a liquid, the atomic configuration varies continuously due to thermal motion but it retains a certain order at any given moment. In a glass, we have a frozen instantaneous configuration which varies very slowly with time. However, a glass has certain characteristic features which distinguish it from ordinary liquids.

First, not every liquid is capable of forming a glass; only liquids with complex twisted molecules, which crystallize with difficulty, are capable of vitrification. Obviously, the supercooling of a liquid and the formation of a glass is easier when the short-range order in a liquid differs markedly from the short-range order in a crystal, and the transition from liquid to crystal involves the overcoming of a certain potential barrier.

Secondly, the process of crystallization may begin on the cooling of a liquid even if this cooling is rapid. Although in glass formation, this process is frozen at a very early stage, the distribution of the atoms in a glass may be somewhat different from that in the initial liquid.

The problem of glass structure was discussed at a conference on this subject held in 1953 [17]. We shall briefly summarize this discussion.

There are essentially two viewpoints. According to one, proposed first by Zachariasen [18], the structure of silicate glasses, which is similar to the structure of the corresponding silicate crystals, represents a continuous network with the atoms or ions at its mesh points. However, in contrast to a regular crystal lattice, this network is irregular, each cell of the crystal lattice being deformed in the glass network. Therefore, although the nearest neighbors are distributed almost regularly with respect to a given central atom, away from this atom the random deformations of the cells are cumulative so that individual atoms or groups of atoms are found to be randomly oriented and distributed. According to this concept, an atomic network, and therefore a glass, is uniform if we ignore inhomogeneities which are of atomic dimensions and are due to fluctuations of the distances between neighboring atoms.

Another viewpoint is that of A. A. Lebedev [19], which suggests that "crystallites" (small regions with ordered distributions of atoms, approaching the distribution in a crystal) may exist in a glass. These crystallites are separated from one another by amorphous phase. According to the crystallite hypothesis, a glass is not completely uniform but contains inhomogeneities in the form of these crystallites which, although small, are much greater than the dimensions of the atoms.

X-ray diffraction analyses cannot distinguish between these two viewpoints. The x-ray diffraction patterns of glasses are curves with several broad maxima. This broadening may be due to short-range order fluctuations (the irregular network hypothesis) or due to small crystallites.

Theoretical intensity curves, plotted using the small crystallite model or the radial distribution of atoms around a central atom (see § 18), give equally good agreement with the experimental curves. Thus, x-ray structure analysis seems to confirm both these contradictory hypotheses.

The supporters of the crystallite hypothesis cite other physicochemical effects which support the presence of microinhomogeneities in a glass, but we shall not consider this any further since the dispute between the supporters of the crystallite structure and those favoring the irregular network structure has lost its point.

Careful analysis has led to a rapprochement between the two viewpoints. The crystallites cannot be regarded as ideal small crystals with sharp boundaries. A crystallite itself is inhomogeneous, with the highest degree of order in its central region, but with the degree of order decreasing on approach to the boundary of the crystallite, with a smooth transition to the amorphous phase. On the other hand, the amorphous phase itself has some short-range order.

This structure may be described by means of the Zachariasen network if it is assumed that the network is not quite uniform but has regions with less or more order, which are order fluctuations on a scale much greater than that of the interatomic spacing.

This unified point of view can explain all the experimental observations. In some cases, crystallites may be more definite while in others the structure may be closer to a uniform irregular network. The transition from one extreme case to the other is gradual and there is no difference in principle between these extremes.

It is likely that the crystallites gradually disappear on heating so that flowing liquids usually have the structure of a continuous space network but in some cases liquids may contain mobile crystallites, which continuously appear and disappear.

§22. Short-Range and Long-Range Order in Alloys

Another example of a quasi-periodic system is a partly ordered solid alloy. Such alloys may exhibit the short-range order in the distribution of the atoms without any long-range order; in this respect, they are similar to liquids.

In a fully ordered atomic alloy at a sufficiently low temperature, the atoms of the different components follow one another in a strict order so that the atoms of each type form a sublattice extending throughout the crystal. This distribution may be called the long-range order in an alloy similar to the long-range order in a simple crystal where atoms are located at the sites of a geometrically regular lattice.

In an ordered alloy, there is also short-range order because each atom has definite nearest neighbors. For example, in a binary alloy crystallizing in a simple cubic lattice of the AB type, each atom has 6 nearest neighbors, all the neighbors of an A atom being B atoms, and conversely. In a fully ordered alloy, the short-range and long-range order are analogous to the situation in a crystal; the short-range order is given by the number and configuration of the nearest neighbors of a given atom.

The fully ordered state of an alloy represents minimum entropy and therefore it will be disturbed by an increase in temperature. Some A and B atoms will interchange places and, consequently, not all the sites of the A sublattice will be occupied by A atoms; some of these sites will be filled by B atoms.

For simplicity, we shall consider only an alloy AB. Let N be the number of sites in each of the two sublattices and let it be also the number of atoms of each type; assume that N_1 is the number of atoms of each type located at the correct sites, N_2 is the number of A atoms at the B sublattice sites or the number of B atoms at the A sublattice sites. Obviously $N_1 + N_2 = N$. The degree of long-range order may be defined as the ratio

$$\xi = \frac{N_1 - N_2}{N}. \qquad (22.1)$$

When all the atoms are at their correct sites, $N_2 = 0$, $N_1 = N$, and $\xi = 1$, i.e., the order is complete. If $N_1 = 0$, $N_2 = N$, and $\xi = 1$ we again have complete order but now the A and B atoms are interchanged. In the absence of order, atoms of a given type are equally likely to be present in either sublattice so that $N_1 = N_2$ and $\xi = 0$.

Similarly, we can define the degree of short-range order η using instead of the number of atoms in both sublattices the number of nearest neighbors.

Let us assume that z_1 is the number of nearest neighbors of a type other than the central atom, z_2 is the number of atoms of the same type as the central atom, and $z_1 + z_2 = z$ is the coordination number of the lattice. Then

$$\eta = \frac{z_1 - z_2}{z}. \qquad (22.2)$$

When $\eta = 1$ obviously $\xi = \pm 1$, and we have complete short-range and long-range order. For small departures from the fully ordered state, η decreases as ξ^2 and then more slowly than ξ^2; if at some temperature $\xi = 0$, η still remains finite, tending to zero only as $T \to \infty$.

The temperature dependence $\eta(T)$ can be found very roughly in the following way [1]. Let u_{AA}, u_{BB}, and u_{AB} be the potential energies of interaction between the atoms indicated in the subscripts. If a B atom which is a neighbor of an A atom is replaced with an A atom, the energy increases by $u_{AA} - u_{AB}$; if an A atom which is a neighbor of a B atom is replaced with a B atom, the energy increases by $u_{BB} - u_{AB}$. Let us use u to denote half the sum of these two differences:

$$u = \frac{1}{2}(u_{AA} + u_{BB}) - u_{AB}. \qquad (22.3)$$

From Boltzmann's law, we find approximately that

$$\frac{z_2}{z_1} = e^{-\frac{u}{k_B T}} \qquad (22.4)$$

and hence

$$\eta = \frac{1 - e^{-\frac{u}{k_B T}}}{1 + e^{\frac{u}{k_B T}}} = \tanh \frac{u}{2 k_B T}. \qquad (22.5)$$

The degree of long-range order can be calculated, also approximately, in the following way. When two atoms A and B are interchanged in a fully ordered alloy, the energy of the crystal increases by

$$W = 2zu, \qquad (22.6)$$

where u is defined by Eq. (22.3). However, as the number of interchanged atoms increases the energy needed to interchange the next pair decreases, reaching zero when the number of atoms at foreign sites equals the number of atoms at their regular sites. Therefore, we may assume that the change in the energy on interchange of a pair of atoms is proportional to the degree of long-range order

$$\Delta E = \xi W. \tag{22.7}$$

From Boltzmann's law, we again find, in analogy to Eq. (22.4),

$$\frac{N_2}{N_1} = e^{-\frac{\xi w}{k_B T}}, \tag{22.8}$$

and then

$$\xi = \tanh \frac{\xi W}{2k_B T}. \tag{22.9}$$

Since the hyperbolic tangent tends to 1 on increase of ξ, and the initial slope of this tangent at the point $\xi = 0$ is $W/2k_B T$, Eq. (22.9) may have a non-zero solution only if $W/2k_B T \geq 1$, i.e., if

$$T \leqslant \theta = \frac{W}{2k_B} ; \tag{22.10}$$

θ is known as the critical temperature for alloy ordering, or the Curie point, in analogy to the Curie point of ferromagnets, which represents the disappearance of spontaneous magnetization.

The inaccuracy of the proposed theory lies in Eq. (22.7), which effectively assumes that the numbers of nearest neighbors z_1 and z_2 are proportional to the numbers N_1 and N_2, which is incorrect. Therefore, the calculations are only of qualitative value and are given simply to illustrate the observation that the long-range order disappears at a certain temperature in an alloy but the short-range order still remains.

The same occurs in liquids which retain their short-range order in the absence of any long-range order.

§ 23. Thermal Motion in Liquids

In § 17 we have shown by a purely phenomenological approach that the thermal motion in a liquid is essentially of the same nature as that in a solid. We shall now consider the thermal motion in liquids in greater detail from the molecular point of view.

We have already mentioned that low-frequency shear vibrations cannot be supported in a liquid because of its flow properties. Thus of the whole possible spectrum of vibrations only the relatively small contribution of transverse waves with low values of **k** is missing in liquids. Since the total number of degrees of freedom of thermal motion should be 3G, where G is the number of atoms, these "missing" vibrations are replaced by other motions of the atoms. They may be damped vibrations or periodic displacements of atoms from one equilibrium position to another. The frequency of such displacements is considerably lower than the characteristic frequency of the thermal vibrations and it determines the rate of change of the configuration of the atoms in a liquid.

Thus, we obtain the following picture of thermal motion: each atom executes a large number of vibrations about one equilibrium position and then jumps to another equilibrium position at a distance δ, which is of the order of atomic dimensions.

In jumping from one equilibrium position to another, an atom must overcome a potential barrier W and therefore the probability of a jump at a given vibration is, in accordance with Boltzmann's law, $\exp(-W/k_B T)$. Consequently, the lifetime of an atom in its equilibrium position, τ, is related to the average period of thermal vibrations τ_0 by

$$\tau = \tau_0 e^{\frac{W}{k_B T}} .$$

(23.1)

The average velocity of displacement of an atom is

$$w = \frac{\delta}{\tau} = \frac{\delta}{\tau_0} e^{-\frac{W}{k_B T}} ,$$

(23.2)

and the self-diffusion coefficient representing the mixing of atoms is

$$D = \frac{\delta^2}{6\tau} = \frac{\delta^2}{6\tau_0} e^{-\frac{W}{k_B T}} .$$

(23.3)

The above formula has been confirmed by direct experiment.

It follows from Eq. (23.1) that at low temperatures the jump frequency is small but it increases with rising temperature so that the jumps become more and more important in thermal motion. However, at not too high temperatures (far from the critical temperatures), the jumps do not make a large contribution to the specific heat of the liquid, scattering of electrons, etc., because the number of jumps is small. However, the jumps play a decisive role in the diffusion and viscous flow of liquids.

The vibrations of atoms about their equilibrium positions generate elastic waves, which will be considered in the same way as in crystals. In the elastic continuum approximation, there is no difference between a crystal and a liquid and therefore this approximation, which usually disregards the anisotropy, applies more to liquids than to crystals.

In the more exact discrete atomic structure model of a liquid, it is necessary to consider the vibrations of a distorted lattice, which introduces certain characteristic features. However, this should introduce no basic changes into the theory of specific heat and electrical conductivity because even the elastic continuum model gives satisfactory results.

Glasses do not exhibit flow properties and therefore their thermal motion, like that of crystals, consists of longitudinal and transverse elastic waves. In the continuum approximation, there is no difference between a glass and a crystal, but in the discrete model the chracteristic features of the atomic structure of a glass (the presence of crystallites and distortions of the space lattice, see § 21) should be allowed for.

Chapter IV

BAND THEORY FOR THE ONE-DIMENSIONAL
MODEL OF A LIQUID

§24. Mechanism of "Melting" in the One-Dimensional
Model

In Chap. I, we established that the properties of metals and many semi-conductors do not change greatly on melting. These properties are closely related to the band structure of the energy spectrum of the electrons in crystals, which has been proved rigorously for a periodic potential. Therefore, the quantum electron theory of liquid and amorphous conductors should begin from the proof that the band structure of the electron energy spectrum applies to liquids and amorphous substances, i.e., that it applies in the absence of strict periodicity.

This will be done in the present chapter for the simplest case of a one-dimensional chain of atoms. It is assumed that initially the atoms are distributed periodically and, consequently, the electron spectrum has a band structure. Then the long-range order in the distribution of the atoms is destroyed completely but the distances between the nearest neighbors are assumed to change only slightly. Perturbation terms then appear in Schrödinger's equation. It will be shown that these perturbation terms do not destroy the band nature of the spectrum; they simply broaden somewhat the allowed bands.

One may object that there is no one-dimensional liquid in existence. The answer to this objection is that the band structure for crystals was first proved for the one-dimensional model, which — in its later stages of refinement — has been frequently employed to study actual physical phenomena. For example, the one-dimensional model indicated the presence of Tamm's surface levels. Similarly, the one-dimensional model of a liquid can be used to establish qualitatively some, but not all, features of the electron spectrum of real three-dimensional liquids. In the present chapter it is also intended to introduce the more complex three-dimensional model for liquids.

We have shown in Chap. III that although a liquid does not have a regular crystal structure or long-range order in the distribution of its atoms with definite sites, it does have a short-range order, i.e., the relative positions of neighboring atoms are approximately the same. How the long-range order may disappear without the short-range order also disappearing can best be seen in the one-dimensional model.

Let us assume that we have initially a one-dimensional crystal consisting of a chain of atoms at equal distances a from one another. So long as there are only small departures from the long-range order (due to, for example, thermal vibrations), the $(N+1)$-th atom is at a distance $(Na + y)$ from the first atom, where $y \ll a$; the same is true for any number N. When this crystal melts, the distance between each pair of neighboring atoms changes slightly, becoming $a(1 + \varepsilon\gamma)$, where $\varepsilon \ll 1$ is a constant parameter which represents the degreee of (small) departure from the short-range order. The quantity γ is different for each pair of neighboring atoms and is of a random nature similar to experimental errors or to the Brownian motion of a particle, because the small changes in the distances between the atoms due to melting are also of a purely random nature.

We shall assume that there is no expansion or compression on melting, since a uniform deformation of the chain introduces nothing new into the situation. In this case, the average value of γ is zero. It is natural to assume that the probability of a given value of γ is represented by the normal Gaussian distribution. We shall normalize this distribution so that the average square of γ is equal to 1:

$$\overline{\gamma^2} = 1. \tag{24.1}$$

The Gaussian distribution also permits large values of γ, i.e., considerable departures from the short-range order, but their probability is low and such departures will be very infrequent.

The distance of the $(N+1)$-th atom from the first atom after melting is $(Na + y)$, where y is the sum of the N values of γ, multiplied by εa. It follows from Lyapunov's theorem that the distribution of the sum of N random quantities tends to the normal distribution on increase of N. Since $\overline{\gamma} = 0$, and $\overline{\gamma^2} = 1$, we obtain the following probability density for the quantities Γ_N, which are sums of N values of γ (see p. 211 in [2]):

$$f(\Gamma_N) = \frac{1}{2\pi N} e^{-\frac{\Gamma_N^2}{2N}} \quad (N \to \infty). \tag{24.2}$$

Hence, the rms value of y is

$$\sqrt{\overline{y^2}} = \varepsilon a \sqrt{N} \qquad (24.3)$$

(see p. 134 in [2]).

If $N > 1/\varepsilon^2$, $\sqrt{\overline{y^2}}$ becomes larger than the lattice constant, i.e., the probability of finding the $(N+1)$-th atom is smeared out over a region greater than the lattice constant and the long-range order disappears. Thus the accumulation of small departures from the short-range order leads to the complete disappearance of the long-range order at distances a/ε^2. It follows from Eq. (24.3) that a considerable departure from the long-range order occurs only if the number of atoms in the region considered is

$$G \gg \frac{1}{\varepsilon^2} . \qquad (24.4)$$

We shall assume that this number G is very large but finite.

Before melting, the potential energy of an electron in the self-consistent field of a chain of atoms is described by a periodic curve $V(x)$ with a period a. On melting, this curve is distorted in two ways. First, all the minima and maxima are shifted in the horizontal direction (along the x axis) following the displacements of the atoms. We can say that each point on the potential curve is shifted along the x axis by the amount y recently introduced. Second, due to changes in the distances between the nearest atoms, the amplitudes of the minima and maxima change, each point of the curves being shifted vertically along the V axis by a quantity ΔV, which is small because ε is small, and which is random like γ.

It is easily seen that distortions of the second kind do not destroy the periodicity of the potential but they introduce small potential energy corrections to the periodic field, which may be easily dealt with by the standard perturbation theory. Therefore, we shall consider primarily the distortions of the first type which can destroy the periodicity of $V(x)$; at given values of x, they change the potential very greatly, even replacing a minimum by a maximum, and conversely.

However, if the coordinate scale is deformed in the same way as the atomic chain on melting, and like the horizontal displacements of the points on the potential curve, i.e., if we introduce a coordinate ξ by means of the relationship,

$$\frac{d\xi}{dx} = \frac{1}{1 + \varepsilon\gamma} , \qquad (24.5)$$

the potential energy of an electron, accurate to within the distortions of the second kind, is still a periodic function of ξ. Similarly, as in the case of alloys (see p. 255 in [3]), we shall separate the potential V into a part V_0, which is periodic in ξ, and a deviation from periodicity which allows for the distortions of the second kind. Since these distortions are the result of atomic shifts, the second term is small and, in the first approximation, proportional to ε:

$$V(\xi) = V_0(\xi) + \varepsilon V'(\xi). \tag{24.6}$$

Since the constant part of the potential can always be included in V_0, we shall assume that

$$\overline{V'} = 0, \tag{24.7}$$

where the bar denotes averaging over the whole chain.

§25. Wave Equation on a Deformed Coordinate Scale

The method of calculation is based on the solution of the wave equation for an electron on a deformed coordinate scale ξ, in which the potential energy of an electron is an approximately periodic function.

Due to thermal motion, the atoms of a "liquid" chain are displaced with time and, therefore, the potential curve and the ξ coordinate scale change. However, since the motion of heavy particles (atoms) is much slower than the motion of electrons, the problem can be solved in the adiabatic approximation, i.e., we can calculate the wave functions of an electron for a certain instantaneous distribution of the atoms, regarding these atoms as fixed. In this case, we can use the stationary Schrödinger equation, which in the Cartesian coordinate x has the form:

$$-\frac{\hbar^2}{2m} \cdot \frac{d^2\psi}{dx^2} + V(x)\psi = E\psi, \tag{25.1}$$

where, as before, $\hbar$ is Planck's constant divided by 2π; m is the electron mass; E is the total energy of the electron; and ψ is the wave function.

Now we shall change over to the coordinate ξ, using Eq. (24.5). Expanding the resultant operator in powers of ε and taking into account Eq. (24.6), we find that Schrödinger's equation becomes:

$$\hat{H}\psi = E\psi; \quad \hat{H} = \hat{H}_0 + \varepsilon V' + \varepsilon \hat{W} + \varepsilon^2 \hat{w} + \ldots; \tag{25.2}$$

$$\hat{H}_0 = -\frac{\hbar^2}{2m} \cdot \frac{d^2}{d\xi^2} + V_0(\xi); \qquad (25.3)$$

$$\hat{W} = \frac{\hbar^2}{2m}\left(2\gamma \frac{d^2}{d\xi^2} + \frac{d\gamma}{d\xi} \cdot \frac{d}{d\xi}\right); \quad \hat{w} = -\frac{3\hbar^2}{2m}\left(\gamma^2 \frac{d^2}{d\xi^2} + \gamma \frac{d\gamma}{d\xi} \cdot \frac{d}{d\xi}\right).$$
$$(25.4)$$

The operator $\hat{H}_0$ is regarded as the unperturbed state and the remaining terms are perturbations.

Thus we have separated the Hamiltonian into a periodic-field operator, and small terms due to departures from short-range order. This is the reason for introducing the coordinates ξ since such a separation cannot be carried out in the coordinate x.

The operators $\hat{W}$ and $\hat{w}$ are not self-adjoint and this must be allowed for in calculating their matrix elements. The operator $\hat{H}$ is completely self-adjoint if, in calculating the matrix elements, we include the orthogonality factor $dx/d\xi$, since the coordinate ξ is not the usual Cartesian one. Since $V_0(\xi)$ is a periodic function, the solution of the unperturbed equation $\hat{H}_0\psi = E\psi$ is discussed in § 13 as the solution of the problem of an electron in a periodic field. The energy spectrum in the unperturbed case represents a series of allowed quasi-continuous bands consisting of G levels (the same as the number of atoms in the chain) each separated by forbidden bands. The eigenvalues of the energy will be denoted by E^0_{nk}, where the subscript n denotes the number of the band and k is the number of the level in the band. Each eigenvalue corresponds to a Bloch wave function:

$$\psi^0_{nk} = \frac{u_{nk}}{\sqrt{G}} e^{ik\xi}; \qquad (25.5)$$

here u_{nk} is a modulating function with the same period as the lattice and normalized to a unit cell; k has G possible values.

The solution for the perturbed case may be sought in the form of a linear combination of a complete system of eigenfunctions of the unperturbed equation:

$$\varphi(\xi) = \sum_{n,k} c_{nk}\psi^0_{nk}(\xi), \qquad (25.6)$$

where, as is known from the perturbation theory, the expansion coefficients c_{nk} are found from the following system of equations:

$$(E^0_{nk} + \varepsilon V'_{nknk} + \varepsilon W_{nknk} + \varepsilon^2 w_{nknk} - E) c_{nk} +$$

$$+ \sum_{n',\, k' \neq n,\, k} (\varepsilon V'_{nkn'k'} + \varepsilon W_{nkn'k'} + \varepsilon^2 w_{nkn'k'}) c_{n'k'} = 0, \quad (25.7)$$

n and k have all possible values; $V_{nkn'k'}$, $W_{nkn'k'}$, and $w_{nkn'k'}$ are the matrix elements of the corresponding perturbation operators. So far, we have not assumed that the perturbation is small.

The unperturbed eigenfunctions are normalized to unity, so that

$$\int_0^L |\psi^0_{nk}|^2 d\xi = 1, \qquad\qquad (25.8)$$

where $L = Ga$ is the length of the atomic chain in the coordinate ξ.

§26.   Calculation of the Perturbation Matrix Elements

To find whether the method of successive approximations, usually employed in the perturbation theory, is applicable, we shall estimate the magnitudes of the perturbation matrix elements. From the derivation of Eq. (25.7), it follows that the matrix elements should be calculated with orthogonality factors of the zeroth-order approximation functions; in the present case, these factors are equal to unity.

Using Eqs. (25.4) and (25.5), we obtain the following expressions:

$$W_{nkn'k'} = \frac{\hbar^2}{2mG} \int_0^L u^*_{nk} \left[2\gamma (u''_{n'k'} + 2ik'u'_{n'k'} - k'^2 u_{n'k'}) + \quad (26.1) \right.$$

$$\left. + \gamma' (u'_{n'k'} + ik'u_{n'k'}) \right] e^{i(k'-k)\xi} d\xi;$$

$$w_{nkn'k'} = -\frac{3\hbar^2}{2mG} \int_0^L u^*_{nk} \left[\gamma^2 (u''_{n'k'} + 2ik'u'_{n'k'} - k'^2 u_{n'k'}) + \right.$$

$$\left. + \gamma\gamma' (u'_{n'k'} + ik'u_{n'k'}) \right] e^{i(k'-k)\xi} d\xi. \qquad (26.2)$$

The primes of u and γ represent differentiation with respect to ξ.

By definition [Eq. (24.5)], the quantity γ is governed by the distance between neighboring atoms during melting. Consequently, within each unit cell, or more exactly within each segment along the x axis between the centers of two neighboring atoms, γ has a definite, almost constant, value which, however, varies at random from segment to segment.

We shall not commit a large error by replacing γ, γ^2, and γ' in an l-th cell with their average values in that cell: γ_l, γ_l^2, and γ_l'.

Next, dividing the integrals from 0 to L into sums of integrals for unit cells, and taking outside the integral sign the factors γ_l, γ_l^2, and γ_l', which are constant in a cell, we shall rewrite the expressions (26.1) and (26.2) in the form:

$$W_{nkn'k'} = \frac{\hbar^2}{2mG}\left(2\sum_{l=1}^{G}\gamma_l K_l + 4ik'\sum_{l=1}^{G}\gamma_l J_l - \right.$$
$$\left. - 2k'^2\sum_{l=1}^{G}\gamma_l I_l + \sum_{l=1}^{G}\gamma_l' J_l + ik'\sum_{l=1}^{G}\gamma_l' I_l\right), \tag{26.3}$$

$$w_{nkn'k'} = -\frac{3\hbar^2}{2mG}\left(\sum_{l=1}^{G}\gamma_l^2 K_l + 2ik'\sum_{n=1}^{G}\gamma_l^2 J_l - \right.$$
$$\left. - k'^2\sum_{l=1}^{G}\gamma_l^2 I_l + \sum_{l=1}^{G}\gamma_l\gamma_l' J_l + ik'\sum_{l=1}^{G}\gamma_l\gamma_l' I_l\right), \tag{26.4}$$

where

$$I_l = \int_{a(l-1)}^{al} u_{nk}^* u_{n'k'} e^{i(k'-k)\xi}d\xi; \quad J_l = \int_{a(l-1)}^{al} u_{nk}^* u_{nk}' e^{i(k'-k)\xi}d\xi; \\ K_l = \int_{a(l-1)}^{al} u_{nk}^* u_{n'k'}'' e^{i(k'-k)\xi}d\xi. \tag{26.5}$$

Since the expressions for $W_{nkn'k'}$ and $w_{nkn'k'}$ include the random quantities γ_l, these matrix elements cannot be calculated exactly but we can find the distribution of the probability of their values using Lyapunov's theorem to calculate the relevant sums. A product of a random quantity γ_l or γ_l', whose sign may vary, and any other function (for example, K_l, J_l, and I_l) gives a new random quantity whose sign may also vary. Therefore, the terms in all the sums in Eq. (26.3) and the last two sums in Eq. (26.4) can have either sign. The change of sign of the terms in the first three sums in Eq. (26.4) is included in the multipliers $\exp[i(k_1-k)\xi]$ in the expressions for K_l, J_l, and I_l. Consequently, the average values of all the sums and, therefore, of the matrix elements are all zero.

The only exceptions are the first three sums in the expressions for the diagonal matrix elements of the operator $\hat{w}$, whose terms have constant signs. Since the functions u_{nk} are periodic, the integrals of Eq. (26.5) are independent of l when $k = k'$ (i.e., the integrals in the diagonal matrix elements are independent of l); we shall denote them by I, J, and K without subscripts. From the normalization condition for the functions ψ^0_{nk}, given in Eq. (25.8), we find that $I = 1$. To determine J and K, we integrate them by parts, using the periodicity of the functions and the order-of-magnitude relationship $|u'| \simeq |u|/a$. As a result of these operations, we find

$$J = \int_0^a u^*_{nk} u'_{nk} d\xi = -\int_0^a u'^*_{nk} u_{nk} d\xi \simeq 0;$$

$$K = \int_0^a u^*_{nk} u''_{nk} d\xi = -\int_0^a |u'_{nk}|^2 d\xi \simeq -\frac{1}{a^2}.$$

Since γ is, strictly speaking, not constant within a unit cell, J is not exactly equal to zero but it is small compared with I and K. Consequently after normalization of the function γ [Eq. (24.1)], the average value of the diagonal matrix element of the operator $\hat{w}$ is

$$\overline{w_{nknk}} = \frac{3\hbar^2}{2m}\left(\frac{1}{a^2} + k^2\right). \qquad (26.6)$$

We can estimate I_l, J_l, and K_l in the same way as I and K. Because the wave functions with different values of k are orthogonal, the absolute values of I_l and K_l are smaller than I and K. However, J_l is not as small as J. We may assume that $|I_l| = \alpha_{kk'}$, $K_l = \dfrac{\alpha_{kk'}}{a^2}$, $J_l = \dfrac{\beta_{kk'}}{a}$, where $\alpha_{kk'}$ and $\beta_{kk'}$ are factors smaller than unity which depend mainly on the difference $k - k'$. When $k - k'$ tends to zero, $\alpha_{kk'}$ tends to unity and $\beta_{kk'}$ tends to zero; when $k - k'$ tends to π/a, $\alpha_{kk'}$ and $\beta_{kk'}$ both tend to zero.

From Lyapunov's theory, it follows that the rms value of the sum of random alternating-sign quantities with identical distributions is equal to the rms value of each term, multiplied by the square root of the number of the terms [see Eqs. (24.1)-(24.3)]. Therefore, using order-of-magnitude estimates, the rms value of the modulus of a matrix element of the operator $\hat{W}$ is found to be

$$\sqrt{\overline{|W_{nkn'k'}|^2}} = \frac{\hbar^2}{2m\sqrt{G}}\sqrt{\alpha^2_{kk'}\left(\frac{4}{a^4} + 4k'^4 + \frac{k'^2}{a^2}\right) + \beta_{kk'}\left(\frac{1}{a^4} + \frac{16k'^2}{a^2}\right)},$$

$$(26.7)$$

and the rms value of the modulus of a nondiagonal matrix element of the operator $\hat{w}$ is

$$\sqrt{\overline{\left|w_{nkn'k'}\right|^2}} = \frac{3\hbar^2}{2m\sqrt{G}}\sqrt{\alpha_{kk'}^2\left(\frac{1}{a^4}+k'^4+\frac{k'^2}{a^2}\right)+\beta_{kk'}^2\left(\frac{1}{a^4}+\frac{4k'^2}{a^2}\right)}.$$

$$(26.8)$$

The potential V' is a random alternating-sign function of ξ, and its matrix elements $V'_{nkn'k'}$ are estimated in a similar way to the matrix elements of $W_{nkn'k'}$:

$$V'_{nkn'k'} = \frac{1}{G}\int V'u_{nk}^* u_{n'k'}e^{i(k'-k)\xi}d\xi =$$

$$= \frac{1}{G}\sum_{l=1}^{G}\int_{a(l-1)}^{al} V'u_{nk}^* u_{n'k'}e^{i(k'-k)\xi}d\xi.$$

$$(26.9)$$

The rms value of Eq. (26.9) is the average value of the sum of alternating-sign random quantities (integrals for each unit cell) and is given by

$$\sqrt{\overline{\left|V'_{nkn'k'}\right|^2}} = \frac{q_{kk'}}{\sqrt{G}},$$

$$(26.10)$$

where $q_{kk'}$ is a quantity which depends primarily on $k-k'$, and its order of magnitude is the same as the average depth of the potential wells.

The order of magnitude of Eqs. (26.7) and (26.10) is the same compared with G and, therefore, for brevity, we shall combine $\hat{W}$ and V', assuming

$$\hat{W}+V'=\hat{U}.$$

$$(26.11)$$

Comparing the expressions (26.6) and (26.10) with (26.7) and bearing in mind the inequality (26.4), we find that

$$\varepsilon^2\hat{w}_{nknk} \gg \varepsilon\sqrt{\overline{\left|U_{nknk}\right|^2}},$$

$$(26.12)$$

i.e., the diagonal matrix elements of the operator $\varepsilon^2\hat{w}$ are larger than those of the operator $\varepsilon\hat{U}$. This means that the term with ε^2 must be retained in the expansion of Eq. (25.2). On the other hand, there is no need to retain operators with higher powers of ε, because, although the diagonal matrix elements of the operators with even powers of ε do not contain $\sqrt{G}$ in the denominator, they are proportional to higher powers of the small quantity ε.

The number of unperturbed energy levels in each band is G and the order-of-magnitude value of the band width is $\hbar^2/ma^2$. Consequently, the distance between neighboring levels E^0_{nk} in a given band is $\hbar^2/ma^2G$ (order-of-magnitude estimate). Using the fact that $k \leq \pi/a$ and the inequality of Eq. (24.4), we find that a nondiagonal perturbation matrix element [Eq. (26.7) multiplied by ε] is larger than the distance between neighboring unperturbed energy levels. This means that the usual method of successive approximations is inapplicable to the discrete spectrum in the present problem.

The problem may be solved by the relative degeneracy method. Relative degeneracy occurs if an unperturbed energy spectrum can be divided into groups of levels, separated by sufficiently wide gaps, but the distance between the levels in each group may be as small as we please. In our case, each such group of levels is an allowed band and the gaps between them are the forbidden bands.

The generalized perturbation method for the relative degeneracy case [4] is based on the separation of a complete exact system of equations such as Eq. (25.7) into several approximate zeroth-order approximation systems which apply to individual groups of levels and are obtained from the exact system by retaining only the terms with n' = n, assuming that $c_{n'k'}$ are small when n' ≠ n. The sole criterion of the applicability of this method — the smallness of a nondiagonal perturbation matrix element compared with the forbidden band widths — is obviously satisfied because of the smallness of $\varepsilon/\sqrt{G}$.

From now on, we shall omit the subscript n. Because of the inequality of Eq. (26.12), we can neglect εU_{kk} compared with $\varepsilon^2 w_{kk}$. Furthermore, since the expressions (26.7) and (26.8) are of the same order of magnitude, we shall drop $\varepsilon^2 w_{kk'}$ from nondiagonal terms. Consequently, the system of equations in the zeroth-order approximation will be, for each band:

$$(E^0_k + \varepsilon^2 w_{kk} - E)c_k + \sum_{k' \neq k} \varepsilon U_{kk'} c_{k'} = 0. \tag{26.13}$$

§ 27. Calculation of the Energy [5-7]

The usual method of calculating the energy is to find the roots of the determinant of the system (26.13). However, finding the roots of a high-order determinant meets with insuperable difficulties. Therefore, the energy spectrum of our system has to be found by means of the following artificial assumption.

We shall assume that we know all the correct linear combinations of the wave functions for the perturbed problem in the zeroth-order approximation

$$\varphi_\nu = \sum_{k=1}^{G} c_{\nu k} \psi_k \quad \nu = 1, 2, \ldots, G, \qquad (27.1)$$

where ψ_k are wave functions (25.5) of the unperturbed problem, satisfying the equations

$$\hat{H}_0 \psi_k = E_k^0 \psi_k. \qquad (27.2)$$

Then the value of the energy of the perturbed system in the first approximation, E_ν, corresponding to the ν-th linear combination may be calculated from

$$E_\nu = \int \varphi_\nu^* \hat{H} \varphi_\nu d\xi, \qquad (27.3)$$

where the total Schrödinger operator is, according to Eqs. (25.2) and (26.11),

$$\hat{H} = \hat{H}_0 + \varepsilon \hat{U} + \varepsilon^2 \hat{w}. \qquad (27.4)$$

Substituting Eqs. (27.1) and (27.4) into Eq. (27.3), using Eq. (27.2) and the orthonormalized nature of the functions ψ_k, we obtain

$$E_\nu = \sum_{k=1}^{G} |c_{\nu k}|^2 E_k^0 + \sum_{k=1}^{G} |c_{\nu k}|^2 (\varepsilon U_{kk} + \varepsilon^2 w_{kk}) +$$
$$+ \sum_{k \neq k'} c_{\nu k}^* c_{\nu k'} (\varepsilon U_{kk'} + \varepsilon^2 w_{kk'}). \qquad (27.5)$$

From the normalization conditions, it follows that

$$\sum_{k=1}^{G} |c_{\nu k}|^2 = 1, \qquad (27.6)$$

and, consequently, on the average, $c_{\nu k}$ is of the order of $1/\sqrt{G}$.

Omitting explicitly the small terms, and assuming that $E_k^0 + \varepsilon w_{kk} = E_k$, we rewrite Eq. (27.5) in the form:

$$E_\nu = \sum_{k=1}^{G} |c_{\nu k}|^2 E_k + \varepsilon \sum_{k, k'} c_{\nu k}^* c_{\nu k'} U_{kk'}; \qquad (27.7)$$

$c_{\nu k}$ satisfy Eqs. (26.13), which can be written as follows:

$$(E_\nu - E_k)c_{\nu k} = \varepsilon \sum_{k''=1}^{G} U_{kk''}c_{\nu k''}. \qquad (27.8)$$

At first sight, it may seem that the second sum in Eq. (27.7) is the sum of random alternating-sign quantities and that, according to Lyapunov's theory, it is of the order of $1/\sqrt{G}$, as erroneously assumed in [5]. In fact, the product $c^*_{\nu k}c_{\nu k'}$ and $U_{kk'}$ are related by Eq. (27.8).

In order to allow for this relationship we shall separate out in $c^*_{\nu k}$ and $c_{\nu k}$ the term with $U_{kk'}$ in accordance with Eq. (27.8) and with the complex-conjugate equation:

$$\left.\begin{array}{l} c^*_{\nu k} = \dfrac{\varepsilon c^*_{\nu k'}U_{kk'}}{E_\nu - E_k} + \left(c^*_{\nu k} - \dfrac{\varepsilon c^*_{\nu k'}U_{kk'}}{E_\nu - E_k}\right) ; \\[3mm] c_{\nu k'} = \dfrac{\varepsilon c_{\nu k}U_{kk'}}{E_\nu - E_{k'}} + \left(c_{\nu k'} - \dfrac{\varepsilon c_{\nu k}U_{k'k}}{E_\nu - E_{k'}}\right) . \end{array}\right\} \qquad (27.9)$$

In the first approximation, the terms in parentheses do not contain $U^*_{kk'}$ and $U_{k'k}$; the dependence on the latter two quantities may occur only through $c_{\nu k'}$ in higher approximations.

The substitution of Eq. (27.9) into Eq. (27.7), omitting the small term

$$-\frac{\varepsilon^2 c_{\nu k}c_{\nu k'}U_{kk'}U^*_{kk'}}{(E_\nu - E_{k'})(E_\nu - E_k)},$$

gives

$$E_\nu = \sum_{k=1}^{G} |c_{\nu k}|^2 (E_k^0 + \varepsilon^2 w_{kk}) + A\varepsilon^2 + B\varepsilon; \qquad (27.10)$$

$$A = \sum_k \frac{\sum'_{k'} |c_{\nu k'}|^2 |U_{kk'}|^2}{E_\nu - E_k} + \sum_{k'} \frac{\sum'_{k} |c_{\nu k}|^2 U_{kk'}U_{k'k}}{E_\nu - E'_k}; \qquad (27.11)$$

$$B = \sum_{k,\,k'}' \left(c^*_{\nu k} - \frac{\varepsilon c^*_{k'}U^*_{kk'}}{E_\nu - E_k}\right)\left(c_{\nu k'} - \frac{\varepsilon c_{\nu k}U_{k'k}}{E_\nu - E_k}\right)U_{kk'}. \qquad (27.12)$$

On the basis of Eq. (27.6), the first term in Eq. (27.10) is a weighted average of the values $E_k = E_k^0 + \varepsilon^2 w_{kk}$, i.e., the band of values of the first terms for various ν lies within the band of E_k and, in practice, coincides with the latter.

To estimate the order of magnitude of A, we shall prove that the average value of the product of symmetric matrix elements $U_{kk'} U_{k'k}$ is always positive. For this purpose, we expand the operator $\hat{U}$ into self-adjoint $\hat{U}'$ and anti-self-adjoint $\hat{U}''$ terms:

$$\hat{U} = \hat{U}' + \hat{U}''; \quad U' = \frac{\hbar^2}{2m} \left(2\gamma \frac{d^2}{d\xi^2} + 2\gamma' \frac{d}{d\xi} + \frac{\gamma''}{2} \right) + V'; \left. \right\}$$
$$U'' = -\frac{\hbar^2}{2m} \left(\gamma' \frac{d}{d\xi} + \frac{\gamma''}{2} \right).$$

(27.13)

In this case

$$U_{kk'} U_{k'k} = |U'_{kk'}|^2 - |U''_{kk'}|^2 + U''_{kk'} U'^{*}_{kk'} - U'_{kk'} U''^{*}_{kk'}, \quad (27.14)$$

and the asterisk denotes complex-conjugate quantities.

The last two terms in Eq. (27.14) are on the average equal to zero, because they are products of different random quantities. It can also be easily shown that

$$|U_{kk'}|^2 > |U''_{kk'}|^2.$$

In fact the operator $\hat{U}'$ consists of the same terms as $\hat{U}''$, but one of them has a coefficient 2 and, moreover, contains the terms $\frac{\hbar^2}{m} \gamma \frac{d^2}{d\xi^2}$ and V'. The relative contributions of separate terms in the operators $\hat{U}'$ and $\hat{U}''$ to the matrix elements $U'_{kk'}$ and $U''_{kk'}$ are independent random quantities, i.e., the mean square of the elements is equal to the sum of the mean squares of these contributions. Therefore, the mean square of $U'_{kk'}$, containing a large number of terms, is larger than the mean square of $U''_{kk'}$, which we had to prove.

Thus, in Eq. (27.11), the sums in the numerators have positive terms. Since $U_{kk'}$ is of the order of magnitude of $1/\sqrt{G}$, and $\sum_{k'} |c_{\nu k'}|^2 = 1$, the sums in the numerators are positive and their order of magnitude is the same as that of $1/G$. Consequently, A is independent of G. If E_ν lies at the lower edge of a band, the denominators are negative and the energy level is depressed further; conversely, for upper levels the denominators are positive and these levels are raised. Thus the term $A\varepsilon^2$ broadens the allowed band. In the outer sums of Eq. (27.11), the terms with small denominators, i.e., with $E_\nu \simeq E_k$ or $E_{k'}$, are the dominant ones. However, with this value of E_ν the main contribution to the inner sums comes from a small number of terms with $k' \approx k$, because, in accordance with Eq. (27.8),

$$\frac{c_{\nu k}}{c_{\nu k'}} \approx \frac{E_\nu - E_{k'}}{E_\nu - E_k} .$$

The sum of a small number of random terms has a large relative dispersion and therefore an allowed band does not just expand but its edges become diffuse.

The term εB may be estimated from Lyapunov's theorem; it is a random alternating-sign quantity with a dispersion of the order of $\varepsilon/\sqrt{G}$, i.e., it disappears as $G \rightarrow \infty$.

From Eq. (27.8) it follows that when $E \simeq E_k$ the quantities c_k are large if $k \approx k'$ and therefore the corresponding solution is a wave packet consisting of Bloch's functions centered on k'.

In order to estimate the broadening of the bands, we shall investigate in more detail the expression (27.11), which, by replacing the summation indices, can be written in the form:

$$A = \sum_k \frac{\sum_{k'} |c_{\nu k'}|^2 (|U_{kk'}|^2 + U_{kk'}U_{k'k})}{E_\nu - E_k} . \qquad (27.15)$$

Because of the random nature of $U_{k'k}$ and because of the summation with respect to k', the numerator in Eq. (27.15) varies at random from one value of k to another, without a regular dependence, but the denominator depends strongly on k. Therefore, we replace the numerator by its average with respect to k, which we shall denote by A'/G (because the numerator is of the order of $1/G$) and take outside the summation sign with respect to k. The remaining sum of k terms can be replaced by an integral because the density of levels is high:

$$\sum \frac{1}{E - E_k} = \frac{2Ga}{\pi} \int \frac{dk}{E - E'(k)} . \qquad (27.16)$$

We shall count the energy from the lower edge of the band E_k. We are interested in the position of the lower edge of an allowed band and, therefore, we shall select ν so that the first term in Eq. (27.10) has the lowest possible value. If G is sufficiently large, the smallest value lies as close as we like to the edge of the band E_k so that we may take it to be equal to zero. Then Eq. (27.16) becomes:

$$E = A\varepsilon^2 = \frac{2a\varepsilon^2 A'}{\pi} \int \frac{dk}{E - E'(k)} , \qquad (27.17)$$

and hence we see that $E < 0$. Consequently, the main contribution to the integral in Eq. (27.16) comes from that part of the band which is lower than E'. E_k differs only by a small term from E_k^0, which is the eigenvalue for a crystal, and therefore the relationship between E' and k in the lower part of the band can be found by the effective mass method. In the one-dimensional case

$$2dk = \frac{\sqrt{2m^*}}{\hbar} \cdot \frac{dE'}{\sqrt{E'}}$$

and

$$A \simeq \frac{2A'a\sqrt{2m^*}}{\pi\hbar} \int_0^{E_1} \frac{dE'}{(E - E')\sqrt{E'}} = -\frac{2A'a\sqrt{2m^*}}{\pi\hbar\sqrt{-E}} \tan^{-1}\sqrt{\frac{E_1}{-E}} \; ; (27.18)$$

here m^* is the effective mass; and E_1 lies somewhere near the middle of the band E_k. The integral for the upper part of the band can be ignored. Similarly, we can investigate the upper edge of an allowed band. Then, counting the energy from the upper edge of the band E_k, we find that $E > 0$ and the formulas contain the effective hole mass, all the other estimates remaining as for electrons.

Assuming that $|E| \ll E_1$, we shall substitute into Eq. (27.18)

$$\tan^{-1}\sqrt{\frac{E_1}{-E}} = \frac{\pi}{2} .$$

Substituting next the new simplified expression into Eq. (27.17) and solving it for E, we find

$$E \simeq -\left(\frac{A'\sqrt{2m^*}a}{\hbar}\right)^{2/3} \varepsilon^{4/3} . \tag{27.19}$$

Since, according to Eqs. (26.7) and (27.11) the order-of-magnitude estimate gives $A' \approx \frac{\hbar^4}{2m^{*2}a^4}$, the final value of the band broadening is

$$\Delta E = -E = \frac{\hbar^2}{2m^*a^2} \varepsilon^{4/3} \simeq E_1 \varepsilon^{4/3}, \tag{27.20}$$

where E_I is the width of the allowed band. Because of the approximation made above, this dependence of ΔE on ε is not exact but we can say that it is close to linear. From Eq. (27.20), we can obtain an approximate criterion for the closing up of the forbidden band when the chain "melts":

$$E_g < \varepsilon (E_I + E_{II}),$$ (27.21)

where E_g is the width of the forbidden band; and E_I and E_{II} are the widths of the neighboring allowed bands.

We have considered so far only one isolated allowed band. Employing the usual perturbation theory formulas, we can also deduce the interaction between bands. In the second approximation, we obtain a correction to the energy of the ν-th level in the n-th band

$$E_{n\nu}^{(2)} = \varepsilon^2 \sum_{n' \neq n} \sum_{\nu'} \frac{U_{n\nu n'\nu'} U_{n'\nu'n\nu}}{E_{n\nu} - E_{n'\nu'}},$$ (27.22)

where $U_{n\nu n'\nu'}$ are matrix elements of the operator $\hat{U}$ calculated in conjunction with the functions (27.1) for various bands. We note that within a single band $U_{n\nu n'\nu'} = 0$ if $\nu \neq \nu'$. Using our earlier estimates of the matrix elements $U_{nkn'k'}$, we can easily show that $E_{n\nu}^{(2)}$ is of the order of $\varepsilon^2 E_{I, II}$ and that it is small compared with the correction to the energy given by the first approximation.

The calculations in the present section are also applicable to a one-dimensional model of a disordered alloy, for which $\hat{W} = 0$ and $\hat{U} = V'$. Since, in an alloy with a small number of components, V' can take up only several discrete values in various cells, the energy spectrum of the alloy has certain characteristic features, for example, the energy band splits into several sub-bands depending on the number of components in the alloy.

§28. Comparison with Numerical Calculations

We have made many assumptions in the theory presented in the foregoing sections. Although these assumptions have been logically justified, estimates of the errors introduced are very difficult to make and therefore it is important to compare the results obtained with numerical calculations for one-dimensional models of liquids with finite numbers of cells. The shortcomings of these calculations are the artificiality of the initial model and the limited applicability of the results, but for our purpose they are quite satisfactory.

The first numerical calculations were carried out by Landauer and Helland [8] but they assumed a very short chain of only 50 atoms. A more interesting calculation was carried out by Makinson and Roberts [9], whose work we shall consider in some detail. They used a model similar to the well-known Kronig-Penney model, with equal δ -type potential wells distributed at random; the

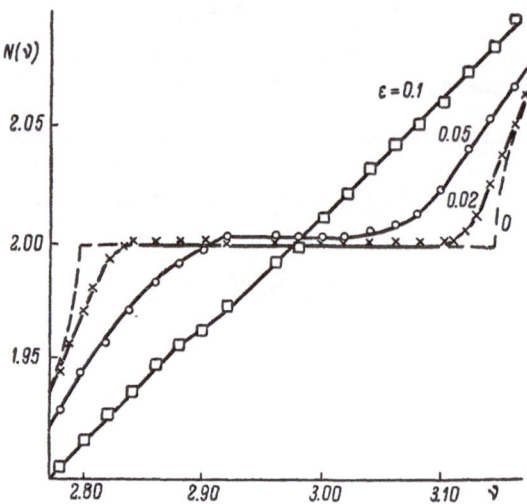

Fig. 18. Number of states $N(\nu)$, per unit length
of a chain, having energy less than ν^2; tight-
binding case, $\lambda = 2$.

number of these wells was 2000. As a further check, calculations were also
carried out for 4000 and 8000 wells. The size of the wells was taken to be
$2\varkappa_0 \hbar^2/2m$, their coordinates were assumed to be x_n, and the electron energy
was taken to be $E = \varkappa_0^2 \nu \hbar^2/2m$; in this notation, Schrödinger's equation becomes:

$$\psi'' + \left[\varkappa_0^2 \nu^2 + 2\varkappa_0 \sum_n \delta(x - x_n) \right] \psi = 0. \qquad (28.1)$$

To solve the problem, we have to find $N(\nu)$, which is the number of
states with the energy less than $\varkappa_0 \nu^2 \hbar^2/2m$ per unit length, i.e., the integral
density of states.

LN(ν), where L is the length of the chain, is equal to the number of
nodes of the function $\psi(x, \nu)$, which has to be calculated using the notation:

$$a_n = \lim_{x \to x_n - 0} \left(\frac{\psi'}{\varkappa_0 \nu \psi} \right) = \tan \chi_n,$$

$$b_n = \lim_{x \to x_n + 0} \left(\frac{\psi'}{\varkappa_0 \nu \psi} \right) = \tan \varphi_n, \qquad (28.2)$$

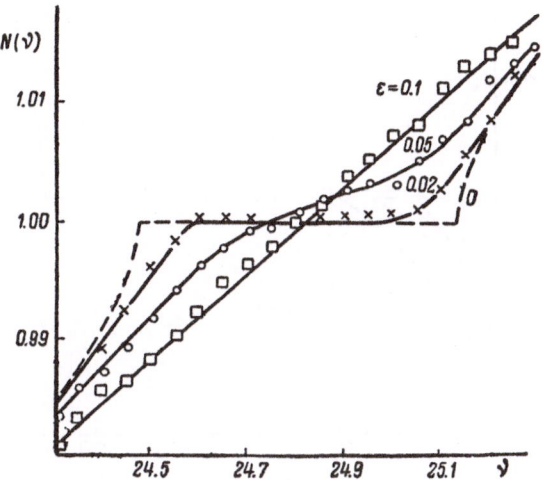

Fig. 19. Number of states N(ν), per unit length of
a chain, having energy less than ν^2; weak-binding
case, $\lambda = 0.125$.

where

$$-\frac{\pi}{2} \leqslant \varphi_n \leqslant \frac{\pi}{2}, \quad u_n = \frac{x_{n+1} - x_n}{a}, \quad a = \overline{x_{n+1} - x_n} \quad \lambda = x_0 a,$$

(28.3)

and integrating Eq. (28.1) once near the point x_n we obtain

$$b_n = a_n - \frac{2}{\nu}, \quad \chi_{n+1} = \varphi_n + \nu \lambda u_n.$$

(28.4)

According to Lax and Phillips [10], the number of nodes in the n-th cell

is the integral part of $\dfrac{\chi_n + \dfrac{\pi}{2}}{\pi}$.

The spectrum is determined by three dimensionless parameters: ν^2,
which is the energy; λ, which is the depth of the potential wells; and u_n,
which is the distance between neighboring atoms. The values of u_n are selected
independently of one another and at random, thus ensuring the absence of long-
range order. For u, the following cutoff parabolic distribution is assumed:

$$dP(u) = \frac{3\sqrt{\pi}}{20\varepsilon} \left[1 - \frac{u-1}{5\varepsilon} \right] du,$$

(28.5)

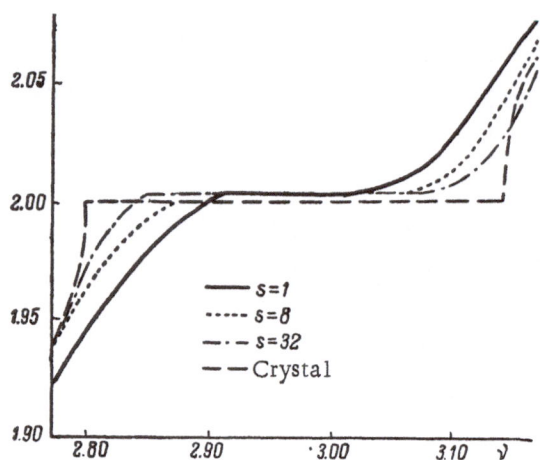

Fig. 20. Influence of smoothing out the potential
on the forbidden band width; $\lambda = 2$, $\varepsilon = 0.05$.

where $\overline{u} = 1$, $1 - \varepsilon\sqrt{5} \leq u \leq 1 + \varepsilon\sqrt{5}$, and ε is the dispersion which has the same value as in the preceding sections.

These calculations have been carried out on an electronic computer for various values of λ and ε. It has been found that the spectrum differs from the periodic Kronig—Penney model only near the forbidden bands. Figure 18 gives the $N(\nu)$ curves for $\lambda = 2$ and $\varepsilon = 0$ (periodic chain), 0.02, 0.05, and 0.1. In Fig. 19 similar curves are given for $\lambda = 0.125$ (the case of weak electron binding). The horizontal parts of the curves represent the forbidden band. When $\varepsilon = 0$, the forbidden band has sharp edges; when ε is increased these edges become more and more diffuse, the allowed bands broaden and gradually fill the forbidden band, and when $\varepsilon = 0.1$ the forbidden band disappears.

In the loose binding case (Fig. 19), the allowed band edges meet already at $\varepsilon = 0.05$ but the density of levels in the forbidden band region, given by the slope of the $N(\nu)$ curves, remains smaller than in the allowed bands. A detailed analysis of the curves in Fig. 18 shows that the broadening of the bands is approximately proportional to ε, but the middle of the forbidden band is displaced slightly to the right at large values of ε, i.e., it is displaced toward higher energies.

Thus, there is complete agreement between the results of the numerical calculations and the results obtained in § 27. The numerical calculations

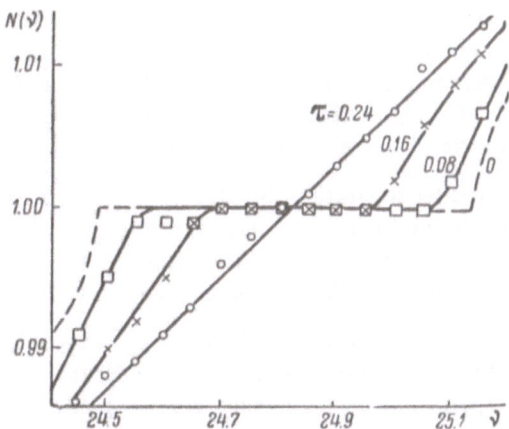

Fig. 21. $N(\nu)$ curve for the weak-binding case
($\lambda = 0.125$) with long-range order retained.

confirm both the broadening of the allowed bands, given approximately by
Eq. (27.20), and the upward displacement of these bands due to the term $\varepsilon^2 w_{kk}$
in Eq. (27.10). Because the atomic chain is finite, the numerical results may
be affected also by the term $B\varepsilon$ in Eq. (27.10), which is proportional to $\varepsilon/\sqrt{G}$,
but this influence is obviously small and cannot affect the basic nature of the
relationships obtained.

Our formulas (27.19) and (27.20) give the average effective band broadening. Since this broadening represents a sum of random quantities, we may
expect the level density at the band edge to obey the Gaussian distribution,
with its "tail" extending further than the average band broadening given by
Eqs. (27.19) and (27.20). In the case of a narrow forbidden band, the Gaussian
distribution "tails" of the neighboring allowed bands meet much earlier than
predicted by Eq. (27.21). In this case, there is no absolutely forbidden band
but only a band with a lower level density. As mentioned above, such a band
is obtained also from the numerical calculations for $\lambda = 0.125$ and $\varepsilon = 0.05$
(Fig. 19). For lower values of ε and $\lambda = 2$, Makinson and Roberts [9] obtained
absolutely forbidden bands, which may be due to their assumption of a cutoff
distribution (28.5) for distances between neighboring atoms. Makinson and
Roberts investigated separately the influence of the disappearance of long-
range and short-range order on the form of the spectrum. In order to allow
separately for the departure from long-range order, they carried out calculations for a model in which each u_n was assumed to be equal to the average

value of u_n for s unit cells located next to the n-th cell in the initial model. It is easily seen that the values of u_n smoothed out in this way differ little for neighboring cells; according to the summation rule for random quantities, $(u_{n+1} - u_n)$ is $\sqrt{s}$ times smaller than in the initial model, although long-range order is disturbed. The curves obtained in this way approach, on increase of s, the curve for a crystal, but their edges remain diffuse.

In order to consider separately the effect of the departure from short-range order, it was assumed that the value of x_n is equally likely to be $(na + \tau)$ or $(na - \tau)$. The curves for $\lambda = 0.125$ are shown in Fig. 21. The allowed bands broaden with increase of τ but the band edges remain fairly sharp, as in crystals.

From a comparison of these investigations, we conclude that in the one-dimensional model the departure from short-range order narrows the forbidden bands, while the departure from long-range order makes the band edges diffuse.

The one-dimensional model of a liquid has been discussed by Rebane [11] and by Eisenschitz and Dean [12]. Rebane considered the joining of the mathematical solutions for individual cells by means of two-dimensional matrices; by investigating the matrix elements, he established the existence of the forbidden bands. This method does not give any other results but it can be extended to the three-dimensional case.

Eisenschitz and Dean derived the electron energy spectrum for a disordered cnain by the tight-binding method. However, as pointed out by Makinson and Roberts [9], they failed to include nondiagonal elements in the energy matrix, which makes their results unreliable.

Chapter V

BAND THEORY FOR THE THREE-DIMENSIONAL
MODEL OF A LIQUID [1]

§29. Mechanism of Melting in the Three-Dimensional Model

In the preceding chapter, we have shown theoretically, for the simplest model of a one-dimensional chain of atoms, that the energy spectrum retains its band structure when the long-range order disappears and there are small departures from the short-range order. In the present chapter, we shall extend this theory to the more realistic case of a three-dimensional model. All the considerations and calculation methods are the same as for the one-dimensional model, therefore we shall repeat our treatment very briefly, pointing out the special features of the three-dimensional case.

Let us assume that we have a crystal with an initially regular distribution of atoms and that the potential energy of an electron in the self-consistent field of the crystal is a periodic function $V(x_1, x_2, x_3)$ of the Cartesian coordinates x_1, x_2, x_3 referred to the crystal. This periodicity is to be understood in the sense that the potential is periodic along certain crystallographic directions which either coincide with the axes x_1, x_2, or x_3 or, in the case of crystals of lower symmetries, are given by linear equations in the coordinate system x_1, x_2, x_3.

X-ray and electron diffraction studies have shown (Chap. III) that a liquid has short-range order in the distribution of its atoms. On melting, the short-range order may either be retained or change suddenly — the number of nearest neighbors of an atom changing (as in the case of water). We shall assume that the short-range order always changes slightly on melting. If a real crystal has a basically different short-range order from a liquid, we should consider instead the melting of a hypothetical crystal with the same short-range order as the liquid. Consequently, we shall assume that each unit cell of a crystal suffers only a small deformation on melting, as a result of which the lengths of the cell edges and angles between them change slightly. The rela-

111

tive deformation of the cell is a small and random quantity ε. For simplicity, we shall assume that on melting a crystal is not subjected to macroscopic deformations, and — in particular — that there is no change of volume. Then all the deformations are equally likely to be positive or negative.

As shown in Chap. IV, the cumulative result of the small deformations of the cells, found from the law of summation of random alternating-sign quantities, is the disappearance of the long-range order at distances of the order of $1/\varepsilon^2$ unit cells. In the three-dimensional case, this means that not only the probability of finding an atom is smeared over a region larger than a unit cell, but also that cells far apart are shifted relative to one another by arbitrary angles.

On melting, the potential field $V(x_1, x_2, x_3)$ suffers two types of distortion due to the deformation of the lattice. First, all the maxima and minima of V are displaced in space parallel to the atomic displacements. Secondly, the value of V changes by a small amount ΔV, due to small departures from the short-range order.

As in the one-dimensional case, the distortions of the second type do not destroy completely the periodicity of the potential but introduce only small corrections to the periodic field, which may be allowed for in the same way as in Chap. IV. If these distortions are not allowed for and if it is assumed that melting simply displaces the points with definite values of the potential parallel to the displacements of the atoms, such an assumption will be equivalent to the hypothesis of deformable ions, which gives a good approximation in the calculation of the interaction between elastic lattice waves and electrons in the theory of metals.

We shall introduce a deformed coordinate system ξ_1, ξ_2, ξ_3 so that the coordinate axes in the liquid pass through the points with the same values of the potential as the corresponding coordinates x_1, x_2, x_3 in a crystal. By suitably selecting the variable scale of the coordinates ξ, we can make the potential quasiperiodic in the ξ coordinate system in the sense that the positions of the maxima and minima of the potential form a regular lattice in the ξ system.

The ξ coordinates are slightly nonorthogonal.

Since the corresponding elements x_α and ξ_α ($\alpha = 1, 2, 3$) may be rotated relative to one another through considerable angles, in the three-dimensional case we use relationships of the (24.5) type between dx_α and $d\xi_\alpha$. However, it is not necessary to know the transformation from the coordinates x to the coordinates ξ; it is sufficient to fix the metric tensor of the ξ coordinate system.

Since each element of a crystal suffers small random deformations on melting, the diagonal and nondiagonal components of the metric tensor should have the form:

$$g_{\alpha\alpha} = 1 + \varepsilon\gamma_{\alpha\alpha}; \quad (g_{\alpha\beta})_{\alpha\neq\beta} = \varepsilon\gamma_{\alpha\beta}; \quad \alpha, \ \beta = 1, \ 2, \ 3; \qquad (29.1)$$

here $\gamma_{\alpha\beta}$ are random functions of the coordinates; they are of the order of unity (some of them may be smaller than unity or equal to zero). In contrast to the one-dimensional case, we cannot normalize all $\gamma_{\alpha\beta}$ exactly since the relationships between them depend on the structure of the liquid. The quantities $\varepsilon\gamma_{\alpha\beta}$ are the components of the tensor of the deformations of individual crystal unit cells which occur on melting.

The functions $\gamma_{\alpha\beta}$ are now defined somewhat differently from the function γ in the one-dimensional case, i.e., using Eq. (29.1) instead of Eq. (24.5), we have

$$\frac{d\xi}{dx} = \frac{1}{\sqrt{1 + \varepsilon\gamma}}. \qquad (29.2)$$

Therefore, if we apply the three-dimensional formulas derived below to the one-dimensional case, we shall obtain results differing from the formulas in Chap. IV by a factor of $\frac{1}{2}$ in the first approximation and $\frac{1}{3}$ in the second approximation.

§30. Wave Equation in a Deformed System of Coordinates

In the three-dimensional case, Schrödinger's equation for an electron is solved in a deformed system of coordinates, in which the potential energy of an electron is a quasiperiodic function. As in the one-dimensional case, the problem is solved in the adiabatic approximation, i.e., for a given instantaneous distribution of the atoms which is assumed to be fixed. We shall use the stationary Schrödinger equation

$$-\frac{\hbar^2}{2m} \nabla^2\psi + V\psi = E\psi. \qquad (30.1)$$

The notation is the same as before. The Laplace operator in a curvilinear oblique system of coordinates is:

$$\nabla^2\psi = \frac{1}{\sqrt{g}} \sum_{\alpha, \beta} \frac{\partial}{\partial\xi_\alpha} \left(g^{\alpha\beta} \sqrt{g} \ \frac{\partial\psi}{\partial\xi_\beta} \right), \qquad (30.2)$$

where g is a determinant composed of covariant components of the metric tensor, whose contravariant components are

$$g^{\alpha\beta} = \frac{G_{\alpha\beta}}{g};$$ (30.3)

$G_{\alpha\beta}$ is the cofactor of the element $g_{\alpha\beta}$ in the determinant, i.e., it is the corresponding minor multiplied by $(-)^{\alpha+\beta}$.

The substitution of Eq. (30.3) into Eq. (30.2) and Eq. (30.2) into Eq. (30.1) and the expansion in powers of ε reduces Schrödinger's equation to the form:

$$\hat{H}\psi = E\psi; \quad \hat{H} = \hat{H}_0 + \varepsilon V' + \varepsilon \hat{W} + \varepsilon^2 \hat{w} + \ldots;$$ (30.4)

$$\hat{H}_0 = -\frac{\hbar^2}{2m} \sum_{\alpha=1}^{3} \frac{\partial^2}{\partial \xi_\alpha^2} + V_0(\xi_1, \xi_2, \xi_3);$$ (30.5)

as in § 25, we have separated the quasiperiodic potential in the ξ system into a periodic part V_0 and a small deviation $\varepsilon V'$, where $\overline{V'} = 0$. Here

$$\hat{W} = \frac{\hbar^2}{2m} \sum_{\alpha=1}^{3} \left[\gamma_{\alpha\alpha} \frac{\partial^2}{\partial \xi_\alpha^2} - \frac{1}{2} \cdot \frac{\partial}{\partial \xi_2} \left(\gamma_{\beta\beta} + \gamma_{\gamma\gamma} - \gamma_{\alpha\alpha} \right) \frac{\partial}{\partial \xi_\alpha} + \right.$$
$$\left. + 2\gamma_{\beta\gamma} \frac{\partial^2}{\partial \xi_\beta \partial \xi_\gamma} + \frac{\partial \gamma_{\beta\gamma}}{\partial \xi_\beta} \cdot \frac{\partial}{\partial \xi_\gamma} + \frac{\partial \gamma_{\beta\gamma}}{\partial \xi_\gamma} \cdot \frac{\partial}{\partial \xi_\beta} \right];$$ (30.6)

α, β, and γ represent cyclic interchange of the indices 1, 2, 3. The operator $\hat{W}$ contains only the alternating-sign terms with the random functions γ_{B_γ} to the first power.

The operator $\hat{w}$ may be separated into two parts:

$$\hat{w} = \hat{w}' + \hat{w}'';$$ (30.7)

$\hat{w}'$ consists of terms with a constant sign, containing squares of $\gamma_{\alpha\beta}$, and is equal to

$$\hat{w}' = -\frac{\hbar^2}{2m} \sum_{\alpha=1}^{3} (\gamma_{\alpha\alpha}^2 + \gamma_{\alpha\beta}^2 + \gamma_{\alpha\gamma}^2) \frac{\partial^2}{\partial \xi_\alpha^2};$$ (30.8)

$\hat{w}''$ consists solely of alternating-sign terms. The expression for $\hat{w}''$ is quite cumbersome and is of no practical importance because $\hat{w}''$ is of the same order of magnitude as $\hat{W}$, also has alternating signs, and is multiplied by ε^2 and not by ε.

The terms $\varepsilon V'$ and the operators $\varepsilon \hat{W}$ and $\varepsilon^2 \hat{W}$ or $\varepsilon^2 \hat{W}'$ are regarded as the perturbation. As in the one-dimensional case, $\hat{W}$ and $\hat{W}'$ are not self-adjoint.

Since $V_0(\xi_1, \xi_2, \xi_3)$ is a periodic function, the solution of the unperturbed equation $\hat{H}_0 \psi = E\psi$ is the well-known solution for an electron in a periodic field. Let us assume that the crystal being considered contains G unit cells, as in the theory of metals. For a considerable departure from the long-range order, the following condition must be satisfied

$$G^{1/3} \gg \frac{1}{\varepsilon^2}. \tag{30.9}$$

The energy spectrum for the unperturbed case consists of several allowed bands, each of which is associated with G Bloch wave functions:

$$\psi_{n\kappa}^0 = \frac{u_{n\kappa}}{\sqrt{G}} e^{i \sum_{\alpha=1}^{3} \kappa_\alpha \xi_\alpha}; \tag{30.10}$$

here n denotes the number of a band, κ_α are coefficients which give the state in a band. For brevity, the three quantities κ_1, κ_2, and κ_3 will be denoted by κ; $u_{n\kappa}$ are modulating functions with the same periodicity as the lattice and normalized to a unit cell.

The solution for the perturbed case will be obtained in the form of a linear combination of the function (30.10)

$$\varphi(\xi_1, \xi_2, \xi_3) = \sum_{n, \kappa} c_{n\kappa} \psi_{n\kappa}^0, \tag{30.11}$$

where the expansion coefficients are found from the following system of equations:

$$(E_{n\kappa}^0 + \varepsilon V'_{n\kappa n\kappa} + \varepsilon W_{n\kappa n\kappa} + \varepsilon^2 w'_{n\kappa n\kappa} - E) c_{n\kappa} +$$
$$+ \sum_{n', \kappa'}' (\varepsilon V'_{n\kappa n'\kappa'} + \varepsilon W_{n\kappa n'\kappa'} + \varepsilon^2 w'_{n\kappa n'\kappa'}) c_{n'\kappa'} = 0. \tag{30.12}$$

The summation with respect to κ is to be carried out for all possible G sets of the three coefficients κ_1, κ_2, κ_3.

The matrix elements $V'_{n\kappa n'\kappa'}$, $W_{n\kappa n'\kappa'}$, and $w'_{n\kappa n'\kappa'}$ are calculated using a weighting function, equal to the orthogonality factor of the zeroth-order approximation functions, i.e., equal to unity. The functions $\psi_{n\kappa}^0$ are normalized to unity for G unit cells.

§31. Calculation of the Perturbation Matrix Elements

According to the expressions (30.6) and (30.10), a matrix element of the operator $\hat{W}$ is in the form:

$$W_{nxn'x'} = \frac{\hbar^2}{2mG} \int u_{nx}^* B_{n'x'} e^{i \sum_{\alpha=1}^{3} (x'_\alpha - x_\alpha) \xi_\alpha} \, d\xi_1 d\xi_2 d\xi_3.$$

(31.1)

An explicit expression for the function $B_{n'x'}$ is somewhat cumbersome, but to determine the order of magnitude of $W_{nxn'x'}$ it is sufficient to note that $B_{n'x'}$ consists of terms of the following type

$$\gamma_{\beta\gamma} u_{n'x'} x'_\alpha x'_\beta; \quad \gamma_{\alpha 3} \frac{\partial u_{n'x'}}{\partial \xi_\alpha} k'_\beta; \quad \gamma_{\alpha\beta} \frac{\partial^2 u_{n'x'}}{\partial \xi_\alpha \partial \xi_\beta};$$

$$\frac{\partial \gamma_{\alpha\beta}}{\partial \xi_\alpha} \cdot \frac{\partial u_{n'x'}}{\partial \xi_\beta} \text{ and } \frac{\partial \gamma_{\alpha\beta}}{\partial \xi_\alpha} u_{n'x'} x'_\beta.$$

The functions $u_{n'x'}$ have the same periodicity as the lattice and the functions $\gamma_{\alpha\beta}$ are of the order of unity but they vary considerably from cell to cell. Therefore, the differentiation of $u_{n'x'}$ and of $\gamma_{\alpha\beta}$ with respect to ξ_α produces a factor whose order of magnitude is $1/a$, where a is the lattice constant. The average value of x'_α is also of the same order of magnitude, $1/a$. From the normalization condition of Eq. (30.13), we find that $u_{n'x'} \simeq a^{-3/2}$. Consequently the order of magnitude of each of the terms in $B_{n'x'}$ is $(1/a^2) u_{n'x'} \simeq a^{-7/2}$. The function $B_{n'x'}$ is of the same order of magnitude and contains a finite number of terms.

To estimate the integral in Eq. (31.1), we shall transform it into a sum of integrals for G unit cells. Within each unit cell, the functions $\gamma_{\alpha\beta}$ and their derivatives have more or less definite values. Therefore, if u_{nx} and $u_{n'x'}$ have similar values, and the factor $\exp\left[i \sum_{\alpha=1}^{3} (x'_\alpha - x_\alpha) \xi_\alpha \right]$ is close to unity, integration over a unit cell reduces, in order of magnitude, to multiplication by a^3; consequently, the order of magnitude of the integral I_l for the l-th unit cell is

$$|I_l| \approx \frac{1}{a^5} \cdot a^3 = \frac{1}{a^2}.$$

(31.2)

In fact, the factors

$$\exp\left[i\sum_{\alpha=1}^{3}(x'_{\alpha}-x_{\alpha})\,\xi_{\alpha}\right]$$

and the inequality of the functions $u_{n\varkappa}$ and $u_{n'\varkappa'}$ make the quantity I_l somewhat smaller, and the larger the difference $(\varkappa-\varkappa')$ the smaller I_l. All the integrals I_l are alternating-sign random quantities. As indicated in detail in Chap. IV, Lyapunov's theorem shows that the sum of alternating-sign random quantities obeys the Gaussian distribution, the average value of this sum is equal to zero and the rms value is proportional to the square root of the number of the terms in the sum. Therefore, summing over all unit cells, we find that the rms value of the matrix element $W_{n\varkappa n'\varkappa'}$ has the following order of magnitude

$$\sqrt{\overline{W^{2}_{n\varkappa n'\varkappa'}}}=\sqrt{G}\,\frac{\hbar^{2}}{2mGa^{2}}=\frac{\hbar^{2}}{2m\sqrt{G}\,a^{2}} \tag{31.3}$$

and decreases as the difference $(\varkappa-\varkappa')$ increases. The matrix elements of the operator $\hat{w}'$ can be found in exactly the same way. The expression (31.2) gives again the order of magnitude of an integral for a unit cell. The integrals for unit cells which occur in the nondiagonal matrix elements $w'_{n\varkappa n'\varkappa'}$ have alternating-signs because of the presence of the factors $\exp\left[i\sum_{\alpha=1}^{3}(x'_{\alpha}-x_{\alpha})\,\xi_{\alpha}\right]$

and, consequently, the order of magnitude of their rms value is again given by Eq. (31.3).

On the other hand, the integrals in a diagonal matrix element $w'_{n\varkappa n\varkappa}$ have constant signs. Placing outside the integral the average values of $\gamma^{2}_{\alpha\beta}$ for a given unit cell and bearing in mind that

$$\left.\begin{array}{c} \int_{1}|u_{n\varkappa}|^{2}\,d\xi_{1}d\xi_{2}d\xi_{3}=1, \quad \int u^{*}_{n\varkappa}\dfrac{\partial u_{n\varkappa}}{\partial\xi_{\alpha}}\,d\xi_{1}d\xi_{2}d\xi_{3}\simeq 0, \\[4mm] \text{and} \\[2mm] \int u^{*}_{n\varkappa}\dfrac{\partial^{2}u_{n\varkappa}}{\partial\xi^{2}_{\alpha}}\,d\xi_{1}d\xi_{2}d\xi_{3}\approx -\dfrac{1}{a^{2}}\,, \end{array}\right\} \tag{31.4}$$

where the integrals are taken over unit cells (see analogous estimates in § 25), we obtain

$$w'_{n\varkappa n\varkappa} \cong \frac{\hbar^2}{2mG} \sum_{\alpha=1}^{3} \sum_{l=1}^{G} (\gamma_{\alpha\alpha}^2 + \gamma_{\alpha\beta}^2 + \gamma_{\alpha\gamma}^2)\left(\frac{1}{a^2} + \varkappa_\alpha^2\right), \qquad (31.5)$$

where $\gamma_{\alpha\beta}$ represent the value of this function for the l-th cell. Carrying out the summation for all values of l, we obtain

$$w'_{n\varkappa n\varkappa} \simeq \frac{\hbar^2}{2m} \sum_{\alpha=1}^{3} (\overline{\gamma_{\alpha\alpha}^2} + \overline{\gamma_{\alpha\beta}^2} + \overline{\gamma_{\alpha\gamma}^2})\left(\frac{1}{a^2} + \varkappa_\alpha^2\right). \qquad (31.6)$$

The bar in Eq. (31.6) represents the average value for all the cells. In a cubic lattice, obviously, $\overline{\gamma^2}_{11} = \overline{\gamma^2}_{22} = \overline{\gamma^2}_{33}$, $\overline{\gamma^2}_{12} = \overline{\gamma^2}_{13} = \overline{\gamma^2}_{23}$, and consequently, assuming that $\sum_{\alpha=1}^{3} \varkappa_\alpha^2 = \varkappa^2$, we have

$$w'_{n\varkappa n\varkappa} \simeq \frac{\hbar^2}{2m} \left(\overline{\gamma_{\alpha\alpha}^2} + 2\overline{\gamma_{\alpha\beta}^2}\right)\left(\frac{3}{a^2} + \varkappa^2\right). \qquad (31.7)$$

Finally, if we assume that $\overline{\gamma^2}_{\alpha\alpha} = \overline{\gamma^2}_{\alpha\beta} = 1$, then

$$w'_{n\varkappa n\varkappa} \simeq \frac{3\hbar^2}{2m} \left(\frac{3}{a^2} + \varkappa^2\right). \qquad (31.8)$$

The matrix element $V'_{n\varkappa n'\varkappa'}$ is estimated in exactly the same way as in the one-dimensional case, i.e., in accordance with Eq. (26.10):

$$\sqrt{\overline{V'^2_{n\varkappa n'\varkappa'}}} = \frac{q_{\varkappa\varkappa'}}{\sqrt{G}}, \qquad (31.9)$$

where $q_{\varkappa\varkappa'}$ is a quantity with the same order of magnitude as the average depth of the potential wells.

Since the matrix elements of $\hat{W}$ and V' are of the same order of magnitude, we shall introduce (as in Chap. IV) the operator

$$\hat{U} = \hat{W} + V'. \qquad (31.10)$$

Comparing the expressions (31.3) and (31.7)-(31.10) and using the inequality (30.9), we find that in Eq. (30.12) we may omit the diagonal element $\varepsilon U_{n\varkappa n\varkappa}$ because it is small compared with $\varepsilon^2 w'_{n\varkappa n\varkappa}$; conversely, the nondiagonal element $\varepsilon^2 w_{n\varkappa n'\varkappa'}$ can be neglected compared with $\varepsilon U_{n\varkappa n'\varkappa'}$.

As in the one-dimensional case, we can show that $\overline{U_{\varkappa\varkappa'}U_{\varkappa'\varkappa}} > 0$.

For this purpose, we separate the operator $\hat{U}$ into self-adjoint $\hat{U}'$ and anti-self-adjoint $\hat{U}''$ parts:

$$\hat{U}' = \frac{\hbar^2}{2m} \sum_{\alpha=1}^{3} \left[\gamma_{\alpha\alpha} \frac{\partial^2}{\partial\xi_\alpha^2} + \frac{\partial\gamma_{\alpha\alpha}}{\partial\xi_\alpha} \cdot \frac{\partial}{\partial\xi_\alpha} + \right.$$
$$\left. + \frac{1}{4} \cdot \frac{\partial^2 (\gamma_{\alpha\alpha} + \gamma_{\beta\beta} + \gamma_{\gamma\gamma})}{\partial\xi_\alpha^2} + 2\gamma_{\beta\gamma} \frac{\partial^2}{\partial\xi_\beta\partial\xi_\gamma} - \frac{1}{2} \cdot \frac{\partial^2\gamma_{\beta\gamma}}{\partial\xi_\beta\partial\xi_\gamma} \right] + V'; \tag{31.11}$$

$$\hat{U}'' = \frac{\hbar^2}{2m} \sum_{\alpha=1}^{3} \left[-\frac{1}{2} \cdot \frac{\partial (\gamma_{\alpha\alpha} + \gamma_{\beta\beta} + \gamma_{\gamma\gamma})}{\partial\xi_\alpha} \cdot \frac{\partial}{\partial\xi_\alpha} - \right.$$
$$\left. -\frac{1}{4} \cdot \frac{\partial^2 (\gamma_{\alpha\alpha} + \gamma_{\beta\beta} + \gamma_{\gamma\gamma})}{\partial\xi_\alpha^2} + \frac{1}{2} \cdot \frac{\partial^2\gamma_{\beta\gamma}}{\partial\xi_\beta\partial\xi_\gamma} + \frac{\partial\gamma_{\beta\gamma}}{\partial\xi_\beta} \cdot \frac{\partial}{\partial\xi_\gamma} + \frac{\partial\gamma_{\beta\gamma}}{\partial\xi_\gamma} \cdot \frac{\partial}{\partial\xi_\beta} \right]. \tag{31.12}$$

In the summation, we use the squares of the random quantities. Therefore, since Eq. (31.12) has terms similar to those in Eq. (31.11) — though in some of them there is a factor of $\frac{1}{2}$, and V' is absent — we may conclude that

$$\overline{|U'_{\varkappa\varkappa'}|^2} > \overline{|U''_{\varkappa\varkappa'}|^2}, \tag{31.13}$$

and therefore

$$\overline{U_{\varkappa\varkappa'}U_{\varkappa'\varkappa}} = \overline{|U'_{\varkappa\varkappa'}|^2} - \overline{|U''_{\varkappa\varkappa'}|^2} > 0. \tag{31.14}$$

If the unperturbed energy of an electron in some band is taken from the lower edge of this band and is proportional to $\varkappa^2$ (the isotropic approximation), we can easily show that the minimum distance between two neighboring levels in the band is

$$\Delta E_{\min} = \frac{2\pi^2\hbar^2}{a^2 G^{2/3}}. \tag{31.15}$$

If we fix the values of $\varkappa_1$ and $\varkappa_2$ and assign to $\varkappa_3$ the neighboring values of 0 and $\pm 2\pi/aG^{1/3}$, we obtain two values of the energy differing by the quantity given in Eq. (31.15). The levels are multiply degenerate, since the number of such levels (order-of-magnitude estimate) does not exceed $G^{2/3}$ and the total number of electron states in a band is G. If the anisotropy is taken into account, the degeneracy is partly lifted so that even more closely spaced energy levels are obtained.

Comparing Eqs. (31.3), (31.9), and (31.15) and using Eq. (30.9), we find that the nondiagonal perturbation matrix elements are larger than the separation between the neighboring levels in the zeroth-order approximation, i.e., we cannot employ the usual perturbation theory but we must use instead, as in the one-dimensional case, the relative degeneracy method. The system of equations in the zeroth-order approximation is completely analogous to Eq. (26.13) and has the form:

$$(E_x - E)\,c_x + \varepsilon \sum_{x'}{}' U_{xx'} c_{x'} = 0, \tag{31.16}$$

where the subscripts n are omitted and

$$E_x = E_x^0 + \varepsilon^2 w'_{xx}. \tag{31.17}$$

§ 32. Calculation of the Energy

The energy may be calculated in exactly the same way as in the one-dimensional case (§ 27). Here we shall note only some special features arising in the three-dimensional approach.

As pointed out in § 30, in the three-dimensional case the summation with respect to one parameter k is replaced by a summation with respect to $\varkappa$, i.e., with respect to all possible sets of the three values $\varkappa_1$, $\varkappa_2$, $\varkappa_3$ which occur in expression (30.10) for the wave function in the zeroth-order approximation. Since the coordinates ξ are not Cartesian, $\varkappa$ is not a wave vector in the physical sense. However, as we are speaking only of functions and spectra in the zeroth-order approximation, we can ignore the non-Cartesian nature of the coordinates ξ and treat $\varkappa$ as a wave vector. The solution of the equation $\hat{H}_0 \psi = E_0 \psi$ in the ξ coordinates has formally the same form as the solution of Schrödinger's equation for a crystal in Cartesian coordinates.

The treatment in § 27, up to Eq. (27.10), may be repeated, replacing the summation indices k and k' with $\varkappa$ and $\varkappa'$, thus obtaining

$$E_y = \sum_x |c_{xy}|^2 (E_x^0 + \varepsilon^2 w_{xx}) + A\varepsilon^2 + B\varepsilon; \tag{32.1}$$

$$A = \sum_x \frac{\sum_{x'}{}' |c_{yx'}|^2 (|U_{xx'}|^2 + U_{xx'}U_{x'x})}{E_y - E_x}; \tag{32.2}$$

$$B = \sum_{x,x'}{}' \left(c_{yx} - \frac{\varepsilon c_{x'} U_{xx'}^*}{E_y - E_x} \right) \left(c_{yx'} - \frac{\varepsilon c_{yx} U_{x'x}}{E_y - E_x} \right) U_{xx'} \tag{32.3}$$

[see the expressions (27.10), (27.12), and (27.15)].

From the relationships (31.3) and (31.9), and the alternating signs of the terms in the sum (32.3), we find that, as in the one-dimensional case, $B \sim 1/\sqrt{G}$; A may be represented in the form:

$$A = \frac{A'}{G} \sum_{\varkappa} \frac{1}{E_\nu - E_\varkappa} . \tag{32.4}$$

The only difference from the one-dimensional case is in the calculation of the sum for all values of $\varkappa$; here, the sum is replaced by an integral over a three-dimensional Brillouin zone

$$A = \frac{A'}{\Omega_\varkappa} \int \frac{d\Omega_\varkappa}{E_\nu - E'(\varkappa)} , \tag{32.5}$$

where $\Omega_\varkappa$ is the volume of this Brillouin zone and $d\Omega_\varkappa$ is an element of this volume. If we select ν so that the first term in (32.1) corresponds to the lower edge of the band $E_\varkappa$ (this edge is taken arbitrarily as the zero energy), then $E = E_\nu$ is negative and close to zero so that the main contribution to the integral in (32.5) comes from that part of the Brillouin zone which lies close to the point of minimum energy. However, in this region, the relationship between E and $\varkappa$ — the latter being treated as a wave vector — can be obtained by the effective mass method. Then,

$$d\Omega_\varkappa = 4\pi\varkappa^2 d\varkappa = \frac{4\pi m^{*3/2}}{\hbar^3} \sqrt{E'} dE'$$

and

$$A \simeq \frac{4\pi A' m^{*3/2} \sqrt{2}}{\Omega_\varkappa \hbar^3} \int_0^{E_1} \frac{\sqrt{E'} dE'}{E - E'} =$$

$$= \frac{8\pi A' m^{*3/2} \sqrt{2}}{\Omega_\varkappa \hbar^3} \left(\sqrt{-E} \tan^{-1} \sqrt{\frac{E_1}{-E}} - \sqrt{E_1} \right) ; \tag{32.6}$$

here, m^* is the effective mass, and E_1 lies somewhere in the middle of the band.

The integral for the upper part of the band is omitted. The upper edge of an allowed band can be treated in the same way.

Assuming that $|E| \ll E_1$, we can drop the term with $\sqrt{-E}$ in Eq. (32.6). Thus, in contrast to the one-dimensional case, in the first approximation A is independent of E. Substituting the simplified expression (32.6) into Eq. (32.1), we obtain

$$E = -\frac{8\pi A' m^{*3/2} \sqrt{2E_1}}{\Omega_x \hbar^3} \, \varepsilon^2. \tag{32.7}$$

The order of magnitude of E_1 is $\hbar^2/2m^* a^2$; according to Eqs. (31.3), (31.10), and (32.2),

$$A' \approx \frac{\hbar^4}{2m^{*2} a^4}, $$

where a is the unit cell dimension, $\Omega_x \approx \pi^{3/2} a^3$, and, consequently, the broadening of the allowed band in the three-dimensional case is

$$\Delta E = -E \simeq \frac{\hbar^2}{2m^* a^2} \, \varepsilon^2 \approx E_I \varepsilon^2, \tag{32.8}$$

where E_I is the allowed band width. The approximate criterion for the disappearance of the forbidden band is

$$E_g \lessapprox \varepsilon^2 (E_I + E_{II}), \tag{32.9}$$

where E_g is the forbidden band width; E_I and E_{II} are widths of the neighboring allowed bands. As in the one-dimensional case, we can allow for the interaction between bands by means of a second-order correction to the energy, described by the same expression as Eq. (27.22):

$$E_{n\nu}^{(2)} = \varepsilon^2 \sum_{n'}{}' \sum_{\nu'} \frac{U_{n\nu n'\nu'} U_{n'\nu' n\nu}}{E_{n\nu} - E_{n'\nu'}}. \tag{32.10}$$

The notation in the above expression is the same as in Eq. (27.22). However, in contrast to the one-dimensional case, due to the special properties of the operator $\hat{U}$, the quantity $E_{n\nu}^{(2)}$ is of the same order of magnitude in terms of ε as the first-approximation correction of Eq. (32.7). The value of $E_{n\nu}^{(2)}$ is still smaller than the term $A\varepsilon^2$ in Eq. (32.1) because the denominator of Eq. (32.10) contains differences in the energies of various bands and, therefore, on the average is greater than the denominator in Eq. (32.2).

The highest value of $E_{n\nu}^{(2)}$ is obtained for a narrow forbidden band. We shall consider $E_{n\nu}^{(2)}$ for the value of ν corresponding to the upper edge of the n-th band. The main contribution to the sum in Eq. (32.10) is made by the band above, for which $E_{n'\nu'} > E_{n\nu}$; on the other hand, using Eq. (31.14), we can show that the numerators in Eq. (32.10) are positive and, therefore, $E_{n\nu}^{(2)} < 0$, i.e., the interband interaction compensates partly for the broadening of the allowed bands, calculated above, and impedes the closing-up of the gap between the bands.

As mentioned at the end of § 27, our calculations may also be applied to disordered alloys, assuming that $\hat{W} = 0$ and $\hat{U} = V'$. The value of A' is, in this case, slighly different: according to Eqs. (31.9) and (32.2), $A' \approx 2q^2_{\varkappa\varkappa}$.

The remark about the possibility of the splitting of an allowed band in a disordered alloy applies also to the three-dimensional case.

We can also make a general remark about the nature of the wave functions in the zeroth-order approximation for a given value of the energy E_ν. These functions are linear combinations of Bloch's functions (30.10) which belong to one band:

$$\varphi_\nu = \sum_\varkappa c_{\nu\varkappa} \psi^0_\varkappa.$$

(32.11)

From Eq. (31.16), it follows that

$$\frac{c_{\nu\varkappa}}{c_{\nu\varkappa'}} \approx \frac{E_\nu - E_{\varkappa'}}{E_\nu - E_\varkappa},$$

(32.12)

and, therefore, the main contribution to Eq. (32.11) is made by terms with $\psi^2_{\varkappa}$, which correspond to energies $E_\varkappa$ lying close to E_ν. The question of the direction of motion of the wave packet represented by Eq. (32.11) is still open and we shall return to this point in the next chapter.

§33. Comparison of the One- and Three-Dimensional Models

After carrying out these calculations for a three-dimensional model of a liquid, which is a much better description of real amorphous substances, we can find how far we were justified in using the one-dimensional model of a liquid, and which of the results obtained from this simplified model are correct and confirmed by the three-dimensional model.

The one-dimensional model gave a band structure of the energy spectrum of the electrons, the broadening of allowed bands, and the spreading of their edges. The same results are obtained with the three-dimensional model. However, the magnitude of the band broadening and its dependence on the degree of departure from short-range order are different for the one- and three-dimensional models, and in this respect the one-dimensional model is unreliable.

In § 27, we have established that in the one-dimensional case the dependence of ΔE (which represents the average broadening of an allowed band) on the degree of departure from order ε is linear, in agreement with the results of the numerical calculations; in the three-dimensional case the dependence is $\Delta E \sim \varepsilon^2$.

The ratio of Eqs. (27.20) and (32.8) gives

$$\frac{\Delta E_{3\text{-dim}}}{\Delta E_{1\text{-dim}}} \approx \varepsilon^{2/3} \ll 1, \tag{33.1}$$

i.e., for the same small values of ε, the band broadening in the three-dimensional case is considerably smaller than that in the one-dimensional model.

This result may be compared with the well-known observation (see, for example, pp. 140 and 185 in [2]) that in the one-dimensional case a local level is formed near a potential well which can be as shallow as we please, while in the three-dimensional case the well must be sufficiently large, i.e., the product of its depth and square of its radius must be large. Any departure from a constant or periodic potential affects more strongly the energy spectrum in the one-dimensional than in the three-dimensional case. This can be easily explained as follows: in the one-dimensional model, any departure from regularity produces a reflection of the electron wave, while in the three-dimensional space the wave may be simply deviated by a local inhomogeneity.

We shall now consider how far we can justify the use of the one-dimensional model in studies of real three-dimensional substances. If we consider the one-dimensional model to be a one-dimensional section through three-dimensional space, or, speaking mathematically, to be the result of eliminating two coordinates from Schrödinger's equation, than a local defect in the one-dimensional model should represent in three-dimensional space not a single impurity atom but a monomolecular layer of such atoms, i.e., a complete crystallographic plane. From this point of view, the one-dimensional model of a liquid represents a three-dimensional structure consisting of parallel crystallographic planes, separated by distances which are random quantities. Such a structure has no application to a real liquid. On the other hand, the surface levels can be considered in the one-dimensional model since the ends of a one-dimensional chain represent crystal faces. Even here, the one-dimensional model is not fully satisfactory because it does not predict a band of surface levels or the effects which are due to surface inhomogeneity.

Thus, we must treat with great caution the results given by the one-dimensional model, especially if the nature of the latter is intimately connected with the method of calculation, as, for example, in the case of the Kronig—Penney model, in the matrix method, and in the numerical calculations.

Such models may lead to inaccurate conclusions about the relationships and details of the electron spectrum in real liquids. Because of this, we have not considered in Chap. IV several one-dimensional theories, regarding as useful only those one-dimensional methods of calculation which can be extended to three dimensions. Among such methods is the deformed coordinate approach.

§34. Changes in the Band Width and Effective Mass when Semiconductors Melt [3]

In all the previous calculations, we have assumed that the coordination number and volume of the substance are not altered on melting. In the present section we shall consider how the energy bands change when real semiconductors melt (we shall consider semiconductors because their properties are particularly sensitive to band structure).

We mentioned in § 2 that two main cases should be distinguished in the melting of a semiconductor.

1. The coordination number increases on melting, the short-range order changes basically, the density increases, and the electrical conductivity becomes metallic. All this is observed in diamond or zinc-blende type semiconductors: germanium, silicon, gallium antimonide, indium antimonide, and other materials for which the coordination number in the solid state is 4 and in the liquid state is 8.

2. The coordination number does not change on melting, the short-range order is disturbed only slightly, the density increases by several percent, and the semiconducting properties are retained. This applies to the melting of oxides and sulfides, for example, Tl_2S, Cu_2Se, Bi_2O_3, V_2O_5, Sb_2S_3.

If the short-range order changes basically on melting, the melt may be compared to a hypothetical crystal which has the same short-range order as the melt. The structures of the energy bands of the melt and of the hypothetical crystal should differ little, although the structures of the bands of the real crystal and the melt may be basically different. The problem of determining the band structure of the melt thus reduces to calculations for a hypothetical crystal having a lattice completely different from that of the real crystal. The energy spectrum and other parameters of this hypothetical crystal have nothing in common with the spectrum and parameters of the real crystal before melting.

According to the calculations of Hund and Mrowka [4], the energy spectrum of a crystal depends strongly on the coordination number z. In particular, in the case of a diamond lattice ($z = 4$), the overlap of the bands representing the s- and p-electrons leads to the formation of a forbidden band and, consequently, the crystal is found to be an insulator or a semiconductor; for higher values of z, a forbidden band is not formed and the crystal exhibits metallic conduction. This explains completely the transition to metallic conduction when semiconductors melt, i.e., case "1", when z increases from 4 to 8. The effective masses and other parameters are completely different in the liquid and solid states.

Case "2" is of greater interest from the point of view of the theory of liquids because here the diffference between a liquid and its crystal reduces to the absence of long-range order and a small increase in volume so that the difference between the energy spectra is much less. We shall now consider specifically this case, which also includes the transitions of semiconductors from the crystalline to the glassy state, described in § 5 and 6.

When the coordination number does not change on melting, the displacements (ΔE) of the energy levels of the electrons can be divided into displacements due to an increase in volume $\Delta_1 E$ and displacements due to a departure from the long-range order without a change in volume $\Delta_2 E$. Assuming both these displacements to be small, we can calculate them independently and then add to obtain

$$\Delta E = \Delta_1 E + \Delta_2 E. \tag{34.1}$$

The calculation of $\Delta_1 E$ presents no difficulties.

The change in volume on melting represents the uniform expansion of a substance, i.e., it simply represents an increase in the lattice constant a. There are a large number of theoretical papers in which the energy spectrum of electrons has been obtained as a function of a, and therefore we may regard the value of the derivative $\partial E / \partial a$ as being known for many crystals; here, E is taken for a definite value of the electron wave vector **k**.

Let us assume that on melting the volume increases in the ratio $1 + \delta$ and that the change in the lattice constant is then $\Delta a \simeq a\delta/3$, i.e.,

$$\Delta_1 E \simeq \frac{a\delta}{3} \cdot \frac{\partial E}{\partial a}. \tag{34.2}$$

The value of $\Delta_2 E$ has been calculated in § 32, but the final expression (32.8) gives only the order of magnitude and cannot yield the individual properties of various semiconductors. The preceding formulas (32.2) and (32.7) are too complicated and do not give $\Delta_2 E$ in terms of measurable values. Therefore, using the results obtained above, we shall estimate $\Delta_2 E$ again, but less rigorously, in terms of quantities which can be calculated or measured for real semiconductors.

In calculating $\Delta_2 E$, we shall bear in mind that when the long-range order is destroyed, each unit cell suffers a small arbitrary deformation. We shall assume that we know the electron energy in a crystal as a function of the wave vector $\mathbf{k}$ and of the components of the deformation tensor $u_{\alpha\beta}$ for a small uniform deformation:

$$E(\mathbf{k},\ u_{\alpha\beta}),$$

where

$$\alpha,\ \beta = 1,\ 2,\ 3.$$

In a liquid, the unit cells are oriented at random with respect to the general coordinate system and they suffer random deformations. In order to obtain the electron energy in a liquid, we have to average out $E(\mathbf{k}, u_{\alpha\beta})$ over all possible directions of $\mathbf{k}$ and all possible values of $u_{\alpha\beta}$. This averaging process can be justified by considering the motion of an electron using the tight-binding approximation. In this case, the electron energy is, in the first approximation, given by the sum of integrals over all unit cells, which gives the required averaging if the cells are not identical. An electron with a given $\mathbf{k}$ moving in a liquid meets cells oriented at random with respect to $\mathbf{k}$, which means that we must average out over all possible directions of $\mathbf{k}$, since the direction of $\mathbf{k}$ with respect to the crystallographic axes of a unit cell is important in these calculations.

First, we shall average out over all directions of $\mathbf{k}$ and obtain an energy which depends only on the absolute value of k and on the deformation components: $E(k, u_{\alpha\beta})$.

The deformations of individual unit cells can be represented in the form:

$$u_{\alpha\beta} = \varepsilon\gamma_{\alpha\beta}, \tag{34.3}$$

where the departure from short-range order is $\varepsilon \ll 1$, and $\gamma_{\alpha\beta}$ are random functions of coordinates, whose average values are zero (the expansion of a melting crystal is allowed for separately), but the rms value is of the order of unity.

Because ε is small, we can expand $E(k, u_{\alpha\beta})$ as a Taylor's series in terms of the deformation components, in a way similar to that used in the theory of elasticity for the free energy of a deformed crystal. We shall consider only cubic crystals since the majority of semiconductors and metals in the molten state have cubic or near-cubic unit cells.

For brevity, we shall use the notation

$$\frac{\partial E}{\partial u_{\alpha\beta}} = E_{\alpha\beta}, \qquad \frac{\partial^2 E}{\partial u_{\alpha\beta}\partial u_{\gamma\delta}} = E_{\alpha\beta\gamma\delta} \qquad (34.4)$$

and, bearing in mind the crystal symmetry, we obtain the following expression, which is similar to that for the energy of an elastically deformed cubic crystal:

$$E(k, u_{\alpha\beta}) = E^0(k) + \varepsilon E_{11} \sum_{\alpha} \gamma_{\alpha\beta} +$$

$$+ \varepsilon^2 \left[\frac{1}{2} E_{1111} \sum_{\alpha} \gamma_{\alpha\alpha}^2 + E_{1122}(\gamma_{11}\gamma_{22} + \gamma_{11}\gamma_{33} + \gamma_{22}\gamma_{33}) + \quad (34.5) \right.$$

$$\left. + 2E_{1212}(\gamma_{12}^2 + \gamma_{13}^2 + \gamma_{23}^2) \right].$$

Averaging out over all possible values of $\gamma_{\alpha\beta}$ and assuming that

$$\overline{\gamma_{\alpha\beta}} = 0, \qquad \overline{\gamma_{\alpha\alpha}^2} = 1, \qquad \overline{(\gamma_{\alpha\beta})^2}_{\alpha\neq\beta} = \gamma, \qquad (34.6)$$

we find

$$E(k, u_{\alpha\beta}) = E^0(k) + \varepsilon^2 \left(\frac{3}{2} E_{1111} + 6\gamma E_{1212} \right). \qquad (34.7)$$

Since

$$E_{1111} = \frac{a^2}{3} \cdot \frac{\partial^2 E}{\partial a^2},$$

we obtain

$$\Delta_2 E = \varepsilon^2 \left(\frac{1}{2} a^2 \frac{\partial^2 E}{\partial a^2} + 6\gamma \frac{\partial^2 E}{\partial u_{12}^2} \right). \qquad (34.8)$$

The change in the forbidden band width on melting can be written in the form:

$$\Delta E_g = \Delta_1 E_g + \Delta_2 E_g + \Delta_3 E_g, \qquad (34.9)$$

where $\Delta_1 E_g$ and $\Delta_2 E_g$ are due to shifts in the energy levels due to the expansion and random deformation of unit cells on melting; they are calculated using Eqs. (34.2) and (34.8), in which E is replaced by E_g. The origin of the third term $\Delta_3 E_g$ is as follows.

The forbidden band width is equal to the difference between the energy minimum in a conduction band and the energy maximum in a filled band. These two extrema may vary with the direction of $\mathbf{k}$, and, on averaging over all directions of $\mathbf{k}$, they are smoothed out, which leads to an increase in the forbidden band width E_g. In order to calculate $\Delta_3 E_g$, it is necessary to know the structure of the energy spectrum of a real crystal. If the band extrema lie at $k = 0$, obviously $\Delta_3 E_g = 0$. In other cases, $\Delta_3 E_g > 0$, but it is small for crystals with more or less close packing.

The change in the allowed band width ΔE_I can be calculated using a formula analogous to (34.9), except that in such a formula $\Delta_3 E_I \leq 0$.

Since the reciprocal of the effective mass is expressed in terms of the derivatives of E with respect to k, it follows from Eqs. (34.1), (34.2), and (34.8) that the change in the reciprocal isotropic (i.e., averaged out for all directions of $\mathbf{k}$) effective mass is

$$\Delta\left(\frac{1}{m^*}\right) \simeq \frac{a\delta}{3} \cdot \frac{\partial}{\partial a}\left(\frac{1}{m^*}\right) + \varepsilon^2\left[\frac{1}{2}a^2\frac{\partial^2}{\partial a^2}\left(\frac{1}{m^*}\right) + 6\gamma\frac{\partial^2}{\partial u_{12}^2}\left(\frac{1}{m^*}\right)\right].$$

(34.10)

We note that the derivatives

$$\frac{\partial E_g}{\partial a}, \quad \frac{\partial^2 E_g}{\partial a^2}, \quad \frac{\partial}{\partial a}\left(\frac{1}{m^*}\right),$$

etc., which occur in the above formulas can be found not only by theoretical calculations for a deformed crystal, but also, and more reliably, from the experimental data. The values of

$$\frac{\partial E_g}{\partial a}, \quad \frac{\partial^2 E_g}{\partial a^2}, \quad \frac{\partial}{\partial a}\left(\frac{1}{m^*}\right), \quad \text{and} \quad \frac{\partial^2}{\partial a^2}\left(\frac{1}{m^*}\right)$$

should be determined from experiments on the uniform compression of a crystal, while

$$\frac{\partial^2 E_g}{\partial u_{12}^2} \quad \text{and} \quad \frac{\partial^2}{\partial u_{12}^2}\left(\frac{1}{m^*}\right)$$

should be found from shear measurements.

The degree of departure from short-range order, ε, is related to the structural diffusion coefficient introduced in § 18: comparing the expressions (18.15) and (24.3), we find that

$$\varepsilon^2 = \frac{2D}{r_1} , \qquad (34.11)$$

where r_1 is the distance between the nearest atoms.

According to Fig. 16, $D \simeq 0.012$ A for liquid tin at temperatures some tens of degrees above the melting point, and this value remains almost constant; r_1 can be determined from the position of the first maximum of the curves in Fig. 15 and this gives $r_1 \simeq 3.5$ A. Consequently, according to Eq. (34.11), $\varepsilon \simeq 0.1$.

When a crystal is subjected to shear, the electron energy levels are displaced much less than under uniform compression or expansion and the terms in Eqs. (34.8) and (34.10) containing γ are not very important. Therefore, we shall not commit a great error by assuming that $\gamma \simeq 1$.

We see that the change in the forbidden band width and in the effective mass on melting or on the formation of the amorphous state may be expressed in terms of the results of independent measurements.

Considering the energy diagrams plotted as a function of the lattice constant, using the theoretical calculations of several workers (for example [4]), we conclude that semiconductors with a coordination number 6 or more as a rule obey the inequalities

$$\frac{\partial E_g}{\partial a} > 0, \quad \frac{\partial^2 E_g}{\partial a^2} < 0. \qquad (34.12)$$

Hence, it follows that $\Delta_1 E_g > 0$ and, probably, $\Delta_2 E_g < 0$; we have established earlier that $\Delta_3 E_g \gtrless 0$. Thus, on melting, the forbidden band width may either increase or decrease; if the expansion on melting is considerable, we would expect the forbidden band width to increase and, conversely, when the change in volume is small, it is more likely that the forbidden band will become narrower.

Unfortunately, the available experimental data are insufficient to make predictions about the change in the effective mass of carriers on melting.

Chapter VI

SCATTERING OF ELECTRONS IN LIQUIDS
BY DEPARTURES FROM LONG-RANGE ORDER
AND BY DEFECTS

§35. Statement of the Problem

In the preceding chapters, we have shown theoretically that if the departure from short-range order is small, and if the long-range order disappears on melting, the energy spectrum of the electrons retains its original band structure.

The calculations were carried out using a deformed system of coordinates ξ_1, ξ_2, ξ_3, such that the main part of the self-consistent potential of an electron in a liquid $V_0(\xi_1, \xi_2, \xi_3)$ is a periodic function of these coordinates, very much like the self-consistent potential in a crystal, which is a periodic function of the Cartesian coordinates x_1, x_2, x_3. It has been shown that in the ξ coordinate system the solution of the wave equation can be obtained from Bloch wave functions of the type:

$$\psi_{\varkappa} = \frac{1}{\sqrt{G}} u_{\varkappa}(\xi_1, \xi_2, \xi_3) e^{i \sum_{\alpha=1}^{3} \varkappa_{\alpha}\xi_{\alpha}}, \tag{35.1}$$

where $u_{\varkappa}(\xi_1, \xi_2, \xi_3)$ is a function of the same periodicity as $V(\xi_1, \xi_2, \xi_3)$.

The stationary solution is a wave packet consisting of functions (35.1). Thus, the disappearance of the long-range order does not alter basically the energy spectrum of the electrons or the nature of their wave functions.

However, the disappearance of strict periodicity in the distribution of the atoms should give rise to additional scattering of the electrons, a reduction in their mean free path, and an increase in the electrical resistance. As mentioned at the end of § 1, the resistance of a liquid with electronic conduction is partly due to a departure from the regular distribution of the atoms and partly due to the thermal vibrations of these atoms. We shall call the first

131

component of the resistance the "liquid" resistance and the second component the "phonon" resistance.

Table 1 in § 1 gives the ratios of the electrical conductivities of several metals in the solid and liquid phases near the melting point; experimental values are listed as well as those calculated allowing for the change in the Debye temperature θ on melting (this temperature occurs in the expression for the mean free path). For some metals, the agreement between these calculated ratios of the conductivies and the measured values is quite satisfactory. However, for mercury the measured ratio of the conductivities is 3.4, while the calculated one is 2.23. Consequently, the additional electron scattering mechanism, represented by the liquid resistance, must be important in mercury. It is by no means accidental that this liquid resistance has been discovered in mercury; this metal melts at a low temperature when the lattice vibrations are not so intense. We may assume that in other metals, which melt at higher temperatures, the liquid resistance is masked by the phonon resistance.

The present chapter deals with the calculation of the mean free path of electrons in a liquid, which is restricted by the departures from long-range order and by the liquid scattering of electrons. The value of this path can be used to calculate, in the same way as in the theory of electrical conductivity, the liquid resistance, which should be very important in liquids with electronic conduction at low temperatures. At high temperatures, this resistance may represent only a small correction to the usual phonon resistance but the existence of the liquid resistance is of basic importance.

We have mentioned several times that in a liquid there are two types of departure from the periodic potential: some of these departures are associated with a spatial shift of the potential minima and maxima; others are due to fluctuations in the amplitudes of these maxima and minima. In the calculations described in Chaps. IV and V, we have combined the two types of departure from periodicity in a single operator $\hat{U}$. Here, for the purpose of investigating their dependence on $\mathbf{k}$, it is more convenient to find the electron mean free path separately for departures of the first and second type: the liquid scattering will be understood to represent the electron scattering on departures of the first type, while the electron scattering on departures of the second type includes the scattering on various types of defects. In the present chapter, we shall ignore thermal vibrations of the lattice since the calculations apply, strictly speaking, to a glass at $T = 0$.

The calculation of the electron mean free path in the case of departures from periodicity of the first type (which represent the disappearance of long-range order) meets with the difficulty that in the Cartesian system of coordinates x, the departures from the periodic potential are large and, therefore, we cannot use the perturbation theory employing Bloch's functions as the zeroth-order approximation. In the ξ coordinate system, in which the perturbation is small (Chaps. IV and V), it is difficult to describe the motion of an electron in an external field since even a uniform electric field in the ξ system has variable components.

Therefore, we shall solve the problem of the motion of an electron in a liquid, i.e., in a quasiperiodic potential V_0, using Cartesian coordinates, but in the zeroth-order approximation we shall use not the standard Bloch functions but the Bloch functions of the type:

$$\Psi_k = \frac{1}{\sqrt{G}} U_k(\xi_1, \xi_2, \xi_3) e^{i\mathbf{k} \cdot \mathbf{r}}, \qquad (35.2)$$

where U_k is a quasiperiodic function of the deformed coordinates ξ_1, ξ_2, ξ_3. The functions given by Eq. (35.2) differ considerably from those given by (35.1), because

$$\sum_{\alpha=1}^{3} x_\alpha \xi_\alpha \neq \mathbf{kr},$$

since the coordinates ξ_α are not Cartesian.

Functions similar to those given by Eq. (35.2) were first used by Shubin [2] in a one-dimensional model of a metal but further calculations were carried out by Shubin employing the method of weakly bound electrons, the use of which is not justified, especially in the case of semiconductors.

§36. Discussion of the Electron Motion in a Liquid in Terms of Cartesian Coordinates

We shall show that the quasi-Bloch functions of Eq. (35.2) satisfy an equation which differs from Schrödinger's equation only by small terms of the order ε, which is the degree of departure from long-range order when a crystal melts. We shall consider, for lucidity, the one-dimensional case when the quasi-Bloch function has the form:

$$\Psi_k = \frac{1}{\sqrt{G}} U_k(\xi) e^{ikx}, \qquad (36.1)$$

and the modulating function U_k is a strictly periodic function of ξ.

The substitution of the usual Bloch function $\psi_k = U_k(x)e^{ikx}$ into Schröd-inger's equation for a crystal, i.e., for periodic $V_0(x)$, gives the following equa-tion for the function $U_k(x)$ periodic in x:

$$\frac{d^2U_k}{dx^2} + 2ik\frac{dU_k}{dx} - (k^2 + \tilde{V} - \tilde{E})U_k = 0, \tag{36.2}$$

where, for brevity, we write

$$\tilde{V} = \frac{2mV_0}{\hbar^2}; \quad \tilde{E} = \frac{2mE}{\hbar^2}; \tag{36.3}$$

E is the total energy of the electron. Since nothing is changed by a change of symbol, we may assume that when $V_0(\xi)$ is periodic, the periodic function $U_k(\xi)$ satisfies the equation

$$\frac{d^2U_k}{d\xi^2} + 2ik\frac{dU_k}{d\xi} - (k^2 + \tilde{V} - \tilde{E})U_k = 0. \tag{36.4}$$

However, since ξ is not a Cartesian coordinate, Eq. (36.4) is not, unlike Eq. (36.2), the direct consequence of Schrödinger's equation.

We shall now replace ξ with x in (36.4) assuming, as in Chap. IV, that,

$$\frac{dx}{d\xi} = 1 + \xi\gamma, \tag{36.5}$$

where γ is a random function of the coordinates and is of the order of unity. Consequently, we find that

$$(1+\varepsilon\gamma)^2\frac{d^2U_k}{dx^2} + (1+\varepsilon\gamma)\left(2ik + \varepsilon\frac{d\gamma}{dx}\right)\frac{dU_k}{dx} - \tag{36.6}$$
$$- (k^2 + \tilde{V} - \tilde{E})U_k = 0.$$

Multiplying Eqs. (36.2) and (36.3) by $(\hbar^2/2m\sqrt{G})e^{ikx}$, we obtain, res-pectively, $\hat{H}\psi_k = E_k\psi_k$ and $\hat{H}_0^k\psi_k = E_k\psi_k$, and, comparing Eqs. (36.2) and (36.6), we find that

$$\hat{H}\Psi_k = \hat{H}_0^k\Psi_k + \varepsilon\hat{W}^k\Psi_k + \varepsilon^2\hat{w}^k\Psi_k; \tag{36.7}$$

$$\hat{H}_0^k \Psi_k = \frac{\hbar^2}{2m\sqrt{G}} \Big[(1+\varepsilon\gamma)^2 \frac{d^2 U_k}{dx^2} + (1+\varepsilon\gamma)\Big(2ik+\varepsilon\frac{d\gamma}{dx}\Big)\frac{dU_k}{dx} - $$
$$- (k^2+\widetilde{V})\,U_k \Big] e^{ikx};$$

$$(36.8)$$

$$\hat{W}^k \Psi_k = -\frac{\hbar^2}{2m\sqrt{G}} \Big[2\gamma \frac{d^2 U_k}{dx^2} + \Big(2ik\gamma+\frac{d\gamma}{dx}\Big)\frac{dU_k}{dx} \Big] e^{ikx}; \quad (36.9)$$

and

$$\hat{w}^k \Psi_k = -\frac{\hbar^2}{2m\sqrt{G}} \Big[\gamma^2 \frac{d^2 U_k}{dx^2} + \gamma \frac{d\gamma}{dx}\cdot\frac{dU_k}{dx} \Big] e^{ikx}. \quad (36.10)$$

It is evident that the functions (36.1) do indeed satisfy equations differing only by small terms from Schrödinger's equation. The equation $\hat{H}_0^k \Psi_k = E_k \Psi_k$ may be regarded as the zeroth-order approximation, and the operators $\varepsilon\hat{W}^k$ and $\varepsilon^2\hat{w}^k$ may be regarded as perturbations. Integrating by parts the products $\frac{d^2 U_k}{dx^2} e^{ikx}$ and $\frac{dU_k}{dx} e^{ikx}$ in Eqs. (36.8)-(36.10), we obtain expressions for the operators themselves:

$$\hat{H}_0^k = \frac{\hbar^2}{2m} \Big\{ (1+\varepsilon\gamma)^2 \frac{d^2}{dx^2} - \varepsilon(1+\varepsilon\gamma)\Big(2ik\gamma-\frac{d\gamma}{dx}\Big)\frac{d}{dx} - $$
$$- \Big[\widetilde{V}+ik(1+\varepsilon\gamma)\varepsilon\frac{d\gamma}{dx}+k^2\varepsilon^2\gamma^2\Big]\Big\}; \quad (36.11)$$

$$\hat{W}^k = \frac{\hbar^2}{2m} \Big[-2\gamma \frac{d^2}{dx^2} + \Big(2ik\gamma-\frac{d\gamma}{dx}\Big)\frac{d}{dx}+ik\frac{d\gamma}{dx} \Big]; \quad (36.12)$$

$$\hat{w}^k = \frac{\hbar^2}{2m} \Big[-\gamma^2 \frac{d^2}{dx^2} + \Big(2ik\gamma^2-\gamma\frac{d\gamma}{dx}\Big)\frac{d}{dx}+k^2\gamma^2+2ik\gamma\frac{d\gamma}{dx} \Big].$$

$$(36.13)$$

However, in our applications we have to use the expressions (36.8)-(36.10) directly.

The three-dimensional case can be dealt with in exactly the same way. From Schrödinger's equation, we can derive the following equation for the periodic function $U_k(x_1,\ x_2\ x_3)$:

$$\sum_{\alpha=1}^{3} \frac{\partial^2 U_k}{\partial x_\alpha^2} + 2ik\sum_{\alpha=1}^{3} i_\alpha \frac{\partial U_k}{\partial x_\alpha} - (k^2+\widetilde{V}-\widetilde{E})\,U_k = 0, \quad (36.14)$$

where i_α is a unit vector along the coordinate axis x_α.

Replacing x_α with ξ_α, we obtain an equation completely analogous to the one-dimensional equation (36.4):

$$\sum_{\alpha=1} \frac{\partial^2 U_\mathbf{k}}{\partial \xi_\alpha^2} + 2i\mathbf{k} \sum_{\alpha=1}^3 \mathbf{e}_\alpha \frac{\partial U_\mathbf{k}}{\partial \xi_\alpha} - (k^2 + \widetilde{V} - \widetilde{E}) U_\mathbf{k} = 0, \quad (36.15)$$

where $\mathbf{e}_\alpha$ is a unit vector of the tangent to the coordinate line ξ_α. However, in contrast to the one-dimensional case, the solutions of Eq. (36.15) for $U_\mathbf{k}$ are not strictly periodic functions of ξ_α. Although the vector $\mathbf{k}$ has a constant direction, the vectors $\mathbf{e}_\alpha$ have different directions in different unit cells. However, the directions of $\mathbf{e}_\alpha$ vary slowly from cell to cell and, therefore, $U_\mathbf{k}$ are quasiperiodic functions of ξ_α in the sense that the values of $U_\mathbf{k}$ at the corresponding points of neighboring cells are equal to within a small quantity of the order of ε. Even in cells far apart, the values of $U_\mathbf{k}$ differ only slightly. It follows from Eq. (36.15) that they differ as much as $U_\mathbf{k}$ in the usual Bloch functions, corresponding to the same absolute values but different directions of the vector $\mathbf{k}$. This difference is usually neglected in the theory of metals (see, for example, p. 188 in [3]).

We shall now change from the coordinates ξ to the coordinates x in Eq. (36.15), bearing in mind that

$$\left. \frac{\partial U}{\partial \xi_\alpha} = \sum_{\beta=1}^3 \frac{\partial U}{\partial x_\beta} \cdot \frac{\partial x_\beta}{\partial \xi_\alpha}; \quad \mathbf{e}_\alpha \cdot \mathbf{i}_\gamma = \frac{1}{\sqrt{\sum\limits_{\varkappa=1}^3 \left(\frac{\partial x_\varkappa}{\partial \xi_\alpha}\right)^2}} \cdot \frac{dx_\gamma}{d\xi_\alpha}; \\ \frac{\partial^2 U}{\partial \xi_\alpha^2} = \sum_{\beta=1}^3 \left(\sum_{\gamma=1}^3 \frac{\partial^2 U}{\partial x_\beta \partial x_\gamma} \cdot \frac{\partial x_\beta}{\partial \xi_\alpha} \cdot \frac{\partial x_\gamma}{\partial \xi_\alpha} + \frac{\partial U}{\partial x_\beta} \cdot \frac{\partial^2 x_\beta}{\partial \xi_\alpha^2} \right). \right\}$$

$$(36.16)$$

Consequently, we obtain, altering the order of summation,

$$\sum_{\beta=1}^3 \left\{ \sum_{\gamma=1}^3 \frac{\partial^2 U_\mathbf{k}}{\partial x_\beta \partial x_\gamma} \sum_{\alpha=1}^3 \frac{\partial x_\beta}{\partial \xi_\alpha} \cdot \frac{\partial x_\gamma}{\partial \xi_\alpha} + \frac{\partial U_\mathbf{k}}{\partial x_\beta} \left[\sum_{\alpha=1}^3 \frac{\partial^2 x_\beta}{\partial \xi_\alpha^2} + 2i \sum_{\gamma=1}^3 k_\gamma \times \right. \right.$$

$$\left. \left. \times \sum_{\alpha=1}^3 \frac{1}{\sqrt{\sum\limits_{\varkappa=1}^3 \left(\frac{\partial x_\varkappa}{\partial \xi_\alpha}\right)^2}} \cdot \frac{\partial x_\beta}{\partial \xi_\alpha} \cdot \frac{\partial x_\gamma}{\partial \xi_\alpha} \right] \right\} - (k^2 + \widetilde{V} - E) U_\mathbf{k} = 0.$$

$$(36.17)$$

If we neglect quantities of the order of ε, then, within each unit cell, the ξ system of coordinates may be regarded as a Cartesian system rotated with respect to the system x. Then,

$$\frac{\partial x_\beta}{\partial \xi_\alpha} = \cos(x_\beta, \xi_\alpha), \quad \frac{\partial^2 x_\beta}{\partial \xi_\alpha^2} = 0,$$

$$\sum_{\alpha=1}^{3} \frac{\partial x_\beta}{\partial \xi_\alpha} \cdot \frac{\partial x_\gamma}{\partial \xi_\alpha} = \cos(x_\beta, x_\gamma) = \delta_{\beta\gamma}, \quad \sum_{\varkappa=1}^{3} \left(\frac{\partial x_\varkappa}{\partial \xi_\alpha}\right)^2 = 1,$$

here, $\delta_{\beta\gamma} = 1$ when $\beta = \gamma$ and $\delta_{\beta\gamma} = 0$ when $\beta \neq \gamma$. Including terms of the order of ε, we may assume that

$$\left. \begin{array}{l} \displaystyle\sum_{\alpha=1}^{3} \frac{\partial x_\beta}{\partial \xi_\alpha} \cdot \frac{\partial x_\gamma}{\partial \xi_\alpha} = \delta_{\beta\gamma} + \varepsilon\gamma_{\beta\gamma}, \quad \sum_{\alpha=1}^{3} \frac{\partial^2 x_\beta}{\partial \xi_\alpha^2} = \frac{\varepsilon\gamma_\beta}{a}, \\[3ex] \displaystyle\sum_{\alpha=1}^{3} \frac{1}{\sqrt{\displaystyle\sum_{\varkappa=1}^{3}\left(\frac{\partial x_\varkappa}{\partial \xi_\alpha}\right)^2}} \cdot \frac{\partial x_\beta}{\partial \xi_\alpha} \cdot \frac{\partial x_\gamma}{\partial \xi_\alpha} = \delta_{\beta\gamma} + \varepsilon\Gamma_{\beta\gamma}, \end{array} \right\} \quad (36.18)$$

here, $\gamma_{\beta\gamma}$, γ_β, and $\Gamma_{\beta\gamma}$ are random functions of the coordinates and they are of the order of unity. They are defined somewhat differently from the functions $\gamma_{\beta\gamma}$ in Chap. V but they have the same properties except that the average values of $\gamma_{\beta\gamma}$ and $\Gamma_{\beta\gamma}$ cannot vanish simultaneously; a is the lattice constant.

The substitution of Eq. (36.18) into (36.17) reduces the latter to the form:

$$\hat{H}_0^k \Psi_k = E_k \Psi_k, \quad (36.19)$$

where

$$\hat{H}_0^k \Psi_k = \frac{\hbar^2}{2m\sqrt{G}} \left[\sum_{\beta=1}^{3} \left(\frac{\partial^2 U_k}{\partial x_\beta^2} + 2ik_\beta \frac{\partial U_k}{\partial x_\beta} \right) - (\tilde{V} + k^2) U_k + \right.$$

$$+ \varepsilon \sum_{\beta=1}^{3} \sum_{\gamma=1}^{3} \left(\frac{\partial^2 U_k}{\partial x_\beta \partial x_\gamma} \gamma_{\beta\gamma} + 2ik_\gamma \frac{\partial U_k}{\partial x_\beta} \Gamma_{\beta\gamma} \right) +$$

$$\left. + \frac{\varepsilon}{a} \sum_{\beta=1}^{3} \frac{\partial U_k}{\partial x_\beta} \gamma_\beta \right] e^{i\mathbf{k}\cdot\mathbf{r}}. \quad (36.20)$$

Comparing this with the usual expression for $\hat{H}\Psi_{\mathbf{k}}$, we find that

$$\hat{H}\Psi_{\mathbf{k}} = \hat{H}_0^{\mathbf{k}}\Psi_{\mathbf{k}} + \varepsilon\hat{W}^{\mathbf{k}}\Psi_{\mathbf{k}}, \qquad (36.21)$$

where

$$\hat{W}^{\mathbf{k}}\Psi_{\mathbf{k}} = -\frac{\hbar^2}{2m\sqrt{G}}\left[\sum_{\beta=1}^{3}\sum_{\gamma=1}^{3}\left(\frac{\partial^2 U_{\mathbf{k}}}{\partial x_\beta \partial x_\gamma}\gamma_{\beta\gamma} + 2ik_\gamma\frac{\partial U_{\mathbf{k}}}{\partial x_\beta}\Gamma_{\beta\gamma}\right) + \right.$$

$$\left. + \frac{1}{a}\sum_{\beta=1}^{3}\frac{\partial U_{\mathbf{k}}}{\partial x_\beta}\gamma_\beta\right]e^{i\mathbf{k}\cdot\mathbf{r}}. \qquad (36.22)$$

In contrast to the one-dimensional case, the operator proportional to ε^2 is now absent. In fact, it is included in the operator $\varepsilon\hat{W}^k$ and appears explicitly if the functions $\Gamma_{\beta\gamma}$ are expressed in terms of $\gamma_{\beta\gamma}$, which would be an unnecessary complication. As in the one-dimensional case, integration by parts would give expressions for the operators $\hat{H}_0^{\mathbf{k}}$ and $\hat{W}^{\mathbf{k}}$ themselves, but we shall not do this since, in our applications, we directly use the expressions $\hat{H}_0^{\mathbf{k}}\Psi_{\mathbf{k}}$ and $\hat{W}^{\mathbf{k}}\Psi_{\mathbf{k}}$. We note only that the operators $\hat{H}_0^{\mathbf{k}}$ and $\hat{W}^{\mathbf{k}}$ depend on $\mathbf{k}$.

It is important to remember that the functions (35.2) describe the motion of a quasifree electron in a liquid in exactly the same way as Bloch's functions describe the motion of a quasifree electron in a crystal. The acceleration of a wave packet in an external field is also calculated as in a crystal, and we can use the effective mass method. The relationship between the velocity and energy of an electron, and between the acceleration and the force acting on the electron, may be obtained without using the periodicity of the functions $U_{\mathbf{k}}$, but only their property of slow variation with $\mathbf{k}$ (see pp. 332-336 in [4]). However, because Eqs. (36.14) and (36.15) are identical, the dependence of the quasiperiodic functions $U_{\mathbf{k}}$ on $\mathbf{k}$ is the same as for the usual periodic functions. Consequently, all the features of the Bloch model apply in our case as well.

Since the operators $\hat{H}_0^{\mathbf{k}}$ are not self-adjoint and since they depend on $\mathbf{k}$, we must consider separately the orthogonality of their eigenfunctions $\Psi_{\mathbf{k}}$. For this purpose, we shall write as usual Eq. (36.19) for the functions $\Psi_{\mathbf{k}}$ and $\Psi_{\mathbf{k}'}^*$, multiplying the first of these equations by $\Psi_{\mathbf{k}'}^*$, and the second by $\Psi_{\mathbf{k}}$, integrating over the whole volume, and subtracting one equation from the other. Consequently, we shall obtain:

$$\int\left(\Psi_{\mathbf{k}'}\hat{H}_0^{\mathbf{k}}\Psi_{\mathbf{k}} - \Psi_{\mathbf{k}}\hat{H}_0^{\mathbf{k}'*}\Psi_{\mathbf{k}'}^*\right)d\mathbf{r} = (E_{\mathbf{k}} - E_{\mathbf{k}'})\int\Psi_{\mathbf{k}'}^*\Psi_{\mathbf{k}}d\mathbf{r}. \quad (36.23)$$

Substituting here $\hat{H}_0^{\mathbf{k}} = \hat{H} - \varepsilon \hat{W}^{\mathbf{k}}$, $\hat{H}_0^{\mathbf{k}'^*} = \hat{H}^* - \varepsilon \hat{W}^{\mathbf{k}'^*}$, and using the self-adjoint operator $\hat{H}$, we obtain

$$(E_{\mathbf{k}} - E_{\mathbf{k}'}) \int \Psi_{\mathbf{k}'}^* \Psi_{\mathbf{k}} d\mathbf{r} = -\varepsilon \int \left(\Psi_{\mathbf{k}'} \hat{W}^{\mathbf{k}} \Psi_{\mathbf{k}} - \Psi_{\mathbf{k}} \hat{W}^{\mathbf{k}'^*} \Psi_{\mathbf{k}'} \right) d\mathbf{r}.$$

$$(36.24)$$

The vectors $\gamma_{\beta\gamma}$ and $\Gamma_{\beta\gamma}$ in $\hat{W}$ and the factors $e^{i(\mathbf{k}-\mathbf{k}')\cdot\mathbf{r}}$ make the integrand in the right-hand part of Eq. (36.24) a random alternating-sign function of the coordinates. Such integrals have been calculated in Chaps. IV and V and their rms value is proportional to $1/\sqrt{G}$, where G is the number of unit cells in the region considered. Thus, the right-hand part of (36.24) is of the order of $\varepsilon/\sqrt{G}$, i.e., it is practically equal to zero. It means that if $E_{\mathbf{k}} \neq E_{\mathbf{k}'}$, then

$$\int \Psi_{\mathbf{k}'}^* \Psi_{\mathbf{k}} d\mathbf{r} \approx \frac{\varepsilon}{\sqrt{G}} \simeq 0,$$

$$(36.25)$$

i.e., the functions $\Psi_{\mathbf{k}}$ are practically orthogonal.

§ 37. Solution of the Transport Equation for Electrons in a Liquid

Since the operators $\hat{H}_0^{\mathbf{k}}$ depend on $\mathbf{k}$, it is necessary to deduce anew the equations for the quantum transitions. In particular, the question of the completeness of the system of functions $\Psi_{\mathbf{k}}$ still remains open. Nevertheless, using the similarity between $\Psi_{\mathbf{k}}$ and the usual Bloch functions and bearing in mind that there is a large number, G, of the functions $\Psi_{\mathbf{k}}$, we shall assume that the solution of the secular Schrödinger equation

$$i\hbar \frac{\partial \Psi}{\partial t} = \hat{H} \Psi$$

$$(37.1)$$

may be given sufficiently accurately in the form of the series

$$\Psi(\mathbf{r},\ t) = \sum_{\mathbf{k}} c_{\mathbf{k}}(t) \Psi_{\mathbf{k}}(\mathbf{r}) e^{-\frac{iE_{\mathbf{k}}t}{\hbar}}.$$

$$(37.2)$$

Substituting Eq. (37.2) into (37.1), separating each term of the sum in the right-hand part of $\hat{H}\Psi_{\mathbf{k}}$ into $\hat{H}_0^{\mathbf{k}}\Psi_{\mathbf{k}}$ and $\varepsilon\hat{W}^{\mathbf{k}}\Psi_{\mathbf{k}}$, integrating over the volume, and using Eqs. (36.19) and (36.25), we arrive at the usual equations for the coefficients $c_{\mathbf{k}}$ in the theory of quantum transitions

$$i\hbar\dot{c}\left(\mathbf{k}'\right) = \varepsilon W_{\mathbf{k}'\mathbf{k}}e^{\frac{i}{\hbar}\left(E_{\mathbf{k}'}-E_{\mathbf{k}}\right)t}, \tag{37.3}$$

where $\varepsilon W_{\mathbf{k}'\mathbf{k}}$ is a matrix element of a transition from the state $\mathbf{k}$ to the state $\mathbf{k}'$.

$$W_{\mathbf{k}'\mathbf{k}} = \int \Psi_{\mathbf{k}'}^* \hat{W}^k \Psi_{\mathbf{k}} d\mathbf{r} = \frac{\hbar^2}{2m\sqrt{G}a^2} I_{\mathbf{k}'\mathbf{k}}, \tag{37.4}$$

where

$$I_{\mathbf{k}'\mathbf{k}} = \frac{a^2}{\sqrt{G}} \int U_{\mathbf{k}'} \left[\sum_{\beta=1}^{3} \sum_{\gamma=1}^{3} \left(\frac{\partial^2 U_k}{\partial x_\beta \partial x_\gamma} \gamma_{\beta\gamma} + 2ik_\gamma \frac{\partial U_k}{\partial x_\beta} \Gamma_{\beta\gamma} \right) + \right.$$

$$\left. + \frac{1}{a} \sum_{\beta=1}^{3} \frac{\partial U_k}{\partial x_\beta} \gamma_\beta \right] e^{i(\mathbf{k}-\mathbf{k}')\cdot\mathbf{r}} d\tau. \tag{37.5}$$

Further calculations differ little from the usual theory of the electrical conductivity presented in § 15. The integration of Eq. (37.3) with respect to time gives

$$|c(\mathbf{k}')|^2 = \varepsilon^2 |W_{\mathbf{k}'\mathbf{k}}|^2 \Omega\left(E_{\mathbf{k}'}-E_{\mathbf{k}}\right) = \frac{\varepsilon^2\hbar^4}{4m^2Ga^4} |I_{\mathbf{k}'\mathbf{k}}|^2 \Omega\left(E_{\mathbf{k}'}-E_{\mathbf{k}}\right), \tag{37.6}$$

where

$$\Omega(\eta) = 2\frac{1-\cos\frac{\eta t}{\hbar}}{\left(\frac{\eta}{\hbar}\right)^2}. \tag{37.7}$$

In order to obtain the total change in the distribution function $f(\mathbf{k}', t)$ due to the "liquid" scattering, it is necessary to multiply Eq. (37.6) by the probability of the initial state $f(\mathbf{k})$, to sum over all possible values of $\mathbf{k}$, to multiply by the probability that the state $\mathbf{k}'$ is unoccupied using the Pauli principle, i.e., multiply by $[1-f(\mathbf{k}')]$, and to allow for reverse transitions from the state $\mathbf{k}'$ to the state $\mathbf{k}$. In this case, we obtain:

$$f(\mathbf{k}', t) - f(\mathbf{k}', 0) = \frac{\varepsilon^2\hbar^4}{4m^2Ga^4} \sum_{\mathbf{k}} \Omega\left(E_{\mathbf{k}'}-E_{\mathbf{k}}\right) \times$$

$$\times \left\{ |I_{\mathbf{k}'\mathbf{k}}|^2 f(\mathbf{k})[1-f(\mathbf{k}')] - |I_{\mathbf{k}\mathbf{k}'}|^2 f(\mathbf{k}')[1-f(\mathbf{k})] \right\}. \tag{37.8}$$

We note that here the usual relationships between the matrix elements of the forward and reverse transitions $W_{k'k} = W^*_{kk'}$ no longer apply. However, if we restrict ourselves to the isotropic approximation used in the classical theory of electrical conduction, we shall show below that in Eq. (37.8)

$$|I_{k'k}|^2 \simeq |I_{kk'}|^2. \tag{37.9}$$

Using Eq. (37.9), reducing similar terms, and differentiating Eq. (37.8) with respect to time, we find the change in the distribution function due to the "liquid" scattering

$$\left(\frac{\partial f}{\partial t}\right)_l = \frac{\varepsilon^2 \hbar^4}{4m^2 G a^4} \cdot \frac{\partial}{\partial t} \sum_k |I_{k'k}|^2 \, \Omega \, (E_{k'} - E_k)[f(\mathbf{k}) - f(\mathbf{k'})]. \tag{37.10}$$

We assume that the electron energy depends only on k, but not on the direction of **k.** In the **k** space, we shall introduce a spherical system of coordinates, k, θ, φ, the axis of which has the same direction as **k'**. Instead of the sum with respect to **k** we can write the integral

$$\sum_k F(\mathbf{k}) = \Omega_0 \frac{G}{(2\pi)^3} \int k^2 F(k) \sin \theta \, d\varphi \, d\theta \, dk, \tag{37.11}$$

where Ω_0 is the volume of a unit cell.

The function $\Omega(\eta)$ has a sharp maximum at $\eta = 0$ so that, when integrating with respect to k in the right-hand part of Eq. (37.10) we can assume that $k = k'$ in all factors except Ω and we can take these factors outside the integral sign.

The remaining integral is then

$$\int_0^\infty \Omega \, (E_{k'} - E_k) \, dk \simeq \frac{1}{\frac{dE}{dk}} \int_{-\infty}^\infty \Omega \, (\eta) \, d\eta = \frac{2\pi \hbar t}{\frac{dE}{dk}}, \tag{37.12}$$

and, consequently,

$$\left(\frac{\partial f}{\partial t}\right)_l = \frac{\varepsilon^2 \hbar 3 \Omega_0 k'^2}{16 m a^4 \pi^2 \frac{dE}{dk}} \int_0^\pi d\theta \int_0^{2\pi} d\varphi \, |I_{k'k}|^2 \sin \theta \, [f(k) - f(k')]. \tag{37.13}$$

The transport equation for the electrons in a liquid in the steady-state case is:

$$\frac{\partial f}{\partial t} = \left(\frac{\partial f}{\partial t}\right)_l + \left(\frac{\partial f}{\partial t}\right)_{\nabla\varphi,\ \nabla\mu,\ \nabla T} = 0, \qquad (37.14)$$

where the last term describes the change in the distribution function due to the presence of a potential gradient, a density gradient, or a temperature gradient. For simplicity, we shall assume that the electron density and the temperature in the liquid are constant and that a uniform field E_x is applied along the x axis. Along this axis [see [3], Eq. (33.18)], it is known that

$$\left(\frac{\partial f}{\partial t}\right)_{\nabla\varphi} = -\frac{eE_x}{\hbar} \cdot \frac{\partial f}{\partial x} \qquad (37.15)$$

and if the field is not too strong, the distribution function may be found in the form:

$$f = f_0(E) + f_1(E)k_x, \qquad (37.16)$$

where the second term is only a small correction. We shall substitute Eqs. (37.13) and (37.15) into (37.14), and in Eq. (37.13) we shall replace f with $f_1 k_x$ so that the terms f_0 are canceled; in Eq. (37.15) we can put f_0 instead of f, neglecting the small correction. Finally, we have

$$\frac{\varepsilon^2\hbar^3\Omega_0 k'^2}{16m^2a^4\pi^2\frac{dE}{dk}} \int_0^\pi d\theta \int_0^{2\pi} d\varphi\, |I_{k'k}|^2 \sin\theta f_1 (k_x - k'_x) = \frac{eE_x}{\hbar} \cdot \frac{\partial f_0}{\partial k'_x} \cdot$$

$$(37.17)$$

Let us assume that ϑ and ϑ' are, respectively, the angles between x and k or k'. Consequently, $k_x = \cos\vartheta$ and $k'_x = k' \cos\vartheta'$. The angles ϑ and ϑ' are related by

$$\cos\vartheta = \cos\vartheta'\cos\theta + \sin\theta\sin\vartheta'\sin\varphi. \qquad (37.18)$$

Since, within the framework of the isotropic approximation, $I_{k'k}$ is independent of φ, the second term of the above expression is eliminated in the integration with respect to φ, and the integration of the remaining terms with respect to φ gives the factor 2π. On the other hand,

$$\frac{\partial f_0}{\partial k'_x} = \frac{df_0}{dE} \cdot \frac{dE}{dk'} \cdot \frac{k'_x}{k'}, \qquad (37.19)$$

and, therefore, the factor $k'_x = k' \cos\vartheta$ is canceled in the left-hand and right-hand parts of Eq. (37.17). Solving this equation for f_1, designating

$$I = \int_0^\pi |I_{\mathbf{k'k}}|^2 (1 - \cos\theta) \sin\theta \, d\theta \qquad (37.20)$$

and omitting the prime in the case of k, we obtain

$$f_1 = -\frac{8\pi m^2 a^4}{\varepsilon^2 \hbar^3 \Omega_0 k^3 I} \cdot \frac{eE_x}{\hbar} \cdot \frac{df_0}{dE} \left(\frac{dE}{dk}\right)^2. \qquad (37.21)$$

Comparing this expression with the distribution function for free electrons

$$f = f_0 - \frac{df_0}{dE} eE_x l \frac{v_x}{v} \qquad (37.22)$$

and assuming that

$$\frac{v_x}{v} = \frac{k_x}{k} \quad \text{and} \quad \Omega_0 \simeq a^3,$$

we obtain the mean free path of the electrons in a liquid in the case of scattering on the departures from the long-range order

$$l_l = \frac{8\pi m^2}{\hbar^4 I} \cdot \frac{1}{k^2} \left(\frac{dE}{dk}\right)^2 \frac{a}{\varepsilon^2}. \qquad (37.23)$$

If the electron energy is represented in the form

$$E = \frac{\hbar^2 k^2}{2m^*}, \qquad (37.24)$$

which can always be done for electrons in the conduction band of a semiconductor, the expression (37.23) simplifies and we obtain

$$l_l = \frac{8\pi}{I} \cdot \frac{a}{\varepsilon^2} \left(\frac{m}{m^*}\right)^2, \qquad (37.25)$$

where m* is the effective electron mass in that band.

§ 38. Calculation of the Matrix Element and Estimation

of the Electron Mean Free Path in Liquids

In order to estimate the mean free path for scattering in a liquid l_l, and to establish its dependence on k, i.e., on the electron energy, we must calculate the coefficient $I_{k'k}$ in the expression for the matrix element of the relevant transition.

As before (Chaps. IV and V), we shall split the integral in (37.5) into a sum of integrals for unit cells. The factors $\gamma_{\beta\gamma}$, $\Gamma_{\beta\gamma}$, and γ_β for each unit cell can be replaced by a constant and taken outside the sign of the integral for a given cell. These quantities are random functions of the coordinates and, therefore, the summation of the individual terms of (37.5) in the integral for a unit cell, and of the integrals for all unit cells, is carried out in accordance with the law governing the summation of random quantities, i.e., we add the squares of the terms. Since the cells are oriented in all possible ways with respect to k, the integrals for the unit cells obtained after taking the factors $\gamma_{\beta\gamma}$ and $\Gamma_{\beta\gamma}$ outside the integral sign are also random quantities which are independent of these factors. However, the average of the product of independent random quantities is equal to the product of the averages. Therefore, summation over G unit cells in the expression for the square of the modulus of $I_{k'k}$ gives the product of the following quantities: the average squares of $\gamma_{\beta\gamma}$, $\Gamma_{\beta\gamma}$, and γ_β, the average square of the integral for a unit cell, and the factor G, which cancels with the factor $(G^{-1/2})^2$, as follows from Eq. (37.5). Consequently, we obtain

$$| I_{k'k} |^2 \simeq \sum_{\beta=1} \sum_{\gamma=1} \left(\overline{\gamma_{\beta\gamma}^2} \, \overline{| \chi_{\beta\gamma}^{k'k} |^2} + 4k_\gamma^2 a^2 \overline{\Gamma_{\beta\gamma}^2} \, \overline{| \chi_{\beta\gamma}^{k'k} |^2} \right) + \sum_{\beta=1}^{3} \overline{\gamma_{\beta\gamma}^2} \, \overline{| \chi_\beta^{k'k} |^2} ;$$

$$(38.1)$$

here,

$$\chi_{\beta\gamma}^{k'k} = a^2 \int_1 U_{k'}^* \frac{\partial^2 U_k}{\partial x_\beta \partial x_\gamma} e^{i\mathbf{q}\cdot\mathbf{r}} d\mathbf{r};$$

$$\left. \chi_\beta^{k'k} = a \int_1 U_{k'}^* \frac{\partial U_k}{\partial x_\beta} e^{i\mathbf{q}\cdot\mathbf{r}} d\mathbf{r}; \quad \mathbf{q} = \mathbf{k} - \mathbf{k}'. \right\}$$

$$(38.2)$$

The bar denotes averaging over all unit cells, and the integrals of (38.2) are taken for one unit cell.

In the isotropic approximation, and exactly for cubic crystals, we have

$$\overline{\gamma_{11}^2} = \overline{\gamma_{22}^2} = \overline{\gamma_{33}^2}; \quad \overline{\gamma_{12}^2} = \overline{\gamma_{23}^2} = \overline{\gamma_{13}^2}; \quad \overline{\gamma_1^2} = \overline{\gamma_2^2} = \overline{\gamma_3^2};$$

$$\overline{\left|\chi_{11}^{k'k}\right|^2} = \overline{\left|\chi_{22}^{k'k}\right|^2} = \overline{\left|\chi_{33}^{k'k}\right|^2} \quad \text{etc.} \Bigg\} \tag{38.3}$$

Similar conditions apply also to $\Gamma_{\alpha\beta}$. Consequently, Eq. (38.1) may be rewritten in the form:

$$\overline{\left|I_{k'k}\right|^2} \simeq 3\overline{\gamma_{\beta\beta}^2}\,\overline{\left|\chi_{\beta\beta}^{k'k}\right|^2} + 6\overline{\gamma_{\beta\gamma}^2}\,\overline{\left|\chi_{\beta\gamma}^{k'k}\right|^2} +$$
$$+ 4k^2 a^2 \left(\overline{\Gamma_{\beta\beta}^2} + 2\overline{\Gamma_{\beta\gamma}^2}\right)\overline{\left|\chi_\beta^{k'k}\right|^2} + 3\overline{\gamma_\beta^2}\,\overline{\left|\chi_\beta^{k'k}\right|^2}, \tag{38.4}$$

where the subscripts β and γ are not equal. Since $k = k'$, in the isotropic approximation we have $U_{k'} = U_k$, and the condition (37.9) follows from Eqs. (38.2) and (38.4).

In order-of-magnitude estimates, we may assume that

$$\overline{\gamma_{\beta\beta}^2} = \overline{\gamma_{\beta\gamma}^2} = \overline{\Gamma_{\beta\beta}^2} = \overline{\Gamma_{\beta\gamma}^2} = \overline{\gamma_\beta^2} = 1 \tag{38.5}$$

and, therefore,

$$\overline{\left|I_{k'k}\right|^2} \simeq 3\overline{\left|\chi_{\beta\beta}^{k'k}\right|^2} + 6\overline{\left|\chi_{\beta\gamma}^{k'k}\right|^2} + 3\overline{\left|\chi_\beta^{k'k}\right|^2} + 12k^2 a^2\,\overline{\left|\chi_\beta^{k'k}\right|^2}. \tag{38.6}$$

The integrals χ are of the order of unity, but they depend on k and on the angle θ between $\mathbf{k}$ and $\mathbf{k}'$. To find the reason for this dependence, we shall expand the functions U_k in a triple Fourier series within one unit cell:

$$U_k = \sum_{\nu_1=-\infty}^{\infty} \sum_{\nu_2=-\infty}^{\infty} \sum_{\nu_3=-\infty}^{\infty} A_{\nu_1\nu_2\nu_3}^k\, e^{\frac{2\pi i}{a}\sum_{\alpha=1}^{3}\nu_\alpha\xi_\alpha}. \tag{38.7}$$

Here, we are assuming that a unit cell is a cube of edge a, and that the coordinates ξ, which are Cartesian within one unit cell, are taken along the cube edges. All this is valid to within small quantities of the order of ε. We are neglecting the difference between the values of U_k for different unit cells.

Substituting Eq. (38.7) into (38.2) and carrying out the integration, we obtain

$$\left|\chi_{\beta\gamma}^{k'k}\right| = b_{\beta\gamma}^{k} \prod_{\alpha=1}^{3} \frac{\sin\dfrac{aq_\alpha}{2}}{\dfrac{aq_\alpha}{2}} + B_{\beta\gamma}^{k} \prod_{\alpha=1}^{3} \sin\frac{aq_\alpha}{2} \qquad (38.8)$$

and similar expressions for $\left|\chi_{\beta}^{k'k}\right|$.

Here,

$$b_{\beta\gamma}^{k} = 4\pi^2 a^3 \sum_{\nu_1}\sum_{\nu_2}\sum_{\nu_3} \left|A_{\nu_1\nu_2\nu_3}^{k}\right|^2 \nu_\beta\nu_\gamma; \qquad (38.9)$$

$$B_{\beta\gamma}^{k} = 4\pi^2 \sum_{\nu_1}\sum_{\nu_2}\sum_{\nu_3}{}'\sum_{\mu_1}{}'\sum_{\mu_2}{}'\sum_{\mu_3}{}' \frac{A_{\nu_1\nu_2\nu_3}^{k}A_{\mu_1\mu_2\mu_3}^{k}\mu_\beta\mu_\gamma}{\prod_{\alpha=1}^{3}\dfrac{\pi}{a}(\mu_\alpha - \nu_\alpha)} ; \qquad (38.10)$$

q_α are the components of the vector q along the axes ξ_α. The primes of the summation signs indicate that the summation with respect to μ_1, μ_2, μ_3 is taken for $\mu_1 \neq \nu_1$, $\mu_2 \neq \nu_2$, and $\mu_3 \neq \nu_3$. In the denominator of Eq. (38.8), we neglect q_α compared with $(\pi/\alpha)(\mu_\alpha - \nu_\alpha)$. Similarly, we find the coefficients b_β^{k} and B_β^{k} in the expressions for $\left|\chi_\beta^{k'k}\right|$. We note that, if (38.7) converges sufficiently rapidly, the coefficients $B_{\beta\gamma}^{k}$ are small compared with $b_{\beta\gamma}^{k}$.

The analysis of the expressions of the (38.8) type is, in general, difficult and, therefore, we shall assume that $aq_\alpha/2 \ll 1$. It is undoubtedly valid for electrons in the conduction band of a semiconductor but may also be used for metals since it is very unlikely that all three quantities, q_1, q_2, and q_3, would simultaneously have their maximum values. In this approximation, we can neglect the second term in (38.8). Expanding in series the sines in the first terms and retaining terms of the second order of smallness with respect to aq_α, we obtain

$$\left|\chi_{\beta\gamma}^{k'k}\right| \simeq b_{\beta\gamma}^{k}\left(1 - \frac{a^2}{24}\sum_{\alpha=1}^{3} q_\alpha^2\right) = b_{\beta\gamma}^{k}\left(1 - \frac{a^2q^2}{24}\right) =$$

$$= b_{\beta\gamma}^{k}\left[1 - \frac{a^2k^2}{12}(1 - \cos\theta)\right]. \qquad (38.11)$$

Similar expressions may be obtained for $(\chi_\beta^{k'k})$.

The substitution of Eq. (38.11) into Eq. (38.4) gives, after omitting terms of the fourth degree in ak,

$$\left| I_{\mathbf{k'k}} \right|^2 = b^{\mathbf{k}} \left[1 - \frac{a^2 k^2}{6} (1 - \cos \theta) \right] + b_0^{\mathbf{k}} k^2 a^2, \qquad (38.12)$$

where

$$b^{\mathbf{k}} = 3\overline{\gamma_{\beta\beta}^2} (b_{\beta\beta}^{\mathbf{k}})^2 + 6\overline{\gamma_{\beta\gamma}^2} (b_{\beta\gamma}^{\mathbf{k}})^2 + 3\overline{\gamma_{\beta}^2} (b_{\beta}^{\mathbf{k}})^2; \qquad (38.13)$$

$$b_0^{\mathbf{k}} = 4\Gamma_{\beta\beta}^2 (b_{\beta}^{\mathbf{k}})^2 + 8\Gamma_{\beta\gamma}^2 (b_{\beta}^{\mathbf{k}})^2. \qquad (38.14)$$

The coefficients $b^{\mathbf{k}}$ and $b_0^{\mathbf{k}}$ depend relatively weakly on $\mathbf{k}$ (only via the dependence of $U_{\mathbf{k}}$ and $\mathbf{k}$), and, in the first approximation, we can ignore this dependence. Substituting Eq. (38.12) into Eq. (38.20) and integrating, we obtain

$$I \simeq c + c_1 a^2 k^2, \qquad (38.15)$$

where

$$c = 2b^{\mathbf{k}}, \quad c_1 = 2b_0^{\mathbf{k}} - \frac{4}{9} b^{\mathbf{k}}. \qquad (38.16)$$

According to Eq. (37.25), the dependence of the mean free path on the wave number has the form:

$$l_l = \frac{8\pi}{c + c_1 a^2 k^2} \cdot \frac{a}{\varepsilon^2} \left(\frac{m}{m^*} \right)^2. \qquad (38.17)$$

The coefficients c and c_1 are best obtained from experiments since the calculation of these coefficients involves the use of very approximate theoretical formulas and requires a knowledge of the coefficients in the expansion of the functions $U_{\mathbf{k}}$ as Fourier series.

All these calculations are applicable not only to liquids but also to amorphous semiconductors. In contrast to liquids, amorphous solids may exist at low temperatures when the "liquid" scattering is particularly important.

§ 39. Scattering of Electrons on Defects [5]

In our calculations, we have not allowed for the fact that when atoms are displaced the potential maxima and minima are altered. Moreover, in a liquid, apart from the gradual departure from order in the distribution of the atoms, there are always local defects in the form of holes and other imperfections. This means that the total self-consistent field potential is not a periodic function of the deformed coordinates ξ.

The nature of these departures of the potential from (strict) periodicity is exactly the same as in the case of a crystal having impurities and other defects. We shall refer to all the departures from periodicity in the coordinate system as defects, and our terms will be taken to include imperfections characteristic of the liquid state (holes, scatter of the potential minima and maxima), as well as those characteristic of crystals (impurity atoms, vacant sites, atoms at interstices). The presence of defects is responsible for additional scattering of the electrons, apart from the "liquid" scattering. This additional scattering can be easily calculated by a method used for the resistance of alloys.

As in the theory of the resistance of alloys (p. 225 in [3]), we shall expand the self-consistent field potential of an electron in a liquid into a part which is periodic in the coordinate system and a part representing a departure from the periodic potential:

$$V = V_0 + V''; \tag{39.1}$$

where $V'' = \varepsilon V' + V_D$, V has the same meaning as in Chaps. IV and V, and V_D is the potential of strong local inhomogeneities. Obviously, V'' is a small perturbation. The fluctuations in the value of the potential $\varepsilon V'$ due to small departures from short-range order, are small; in the regions of pronounced inhomogeneities of the quasicrystalline lattice of a liquid, the potential V_D may be large, but since such defects are relatively rare the matrix elements of V'', obtained by integrating over the whole region considered, are relatively small.

Consequently, we may use the perturbation formula (43.11) taken from [3] to alter the electron distribution function to allow for the scattering on defects:

$$\left(\frac{\partial f}{\partial t}\right)_D = \frac{G\Omega_0}{2\pi\hbar} \cdot \frac{k^2}{\dfrac{dE}{dk}} \cdot \frac{df_0}{dE} \chi(E) k_x \int_0^\pi |V(\vartheta)|^2 (1 - \cos\vartheta) \sin\vartheta\, d\vartheta; \tag{39.2}$$

here, ϑ is the angle between electron wave vectors $\mathbf{k}$ and $\mathbf{k'}$. It is assumed that an electric field or a temperature gradient, responsible for the departure of the distribution function from its equilibrium form, is directed along the x axis, so that the distribution function can be represented in the form:

$$f(\mathbf{k}) = f_0(E) - \frac{df_0}{dE} \chi(E) k_x. \tag{39.3}$$

Next,

$$|V(\vartheta)|^2 = \frac{1}{G}\overline{|V_n^{\mathbf{kk'}}|^2}; \quad V_n^{\mathbf{kk'}} = \int_1 V''(r)\, U_{\mathbf{k'}}^* U_{\mathbf{k}} e^{i(\mathbf{k}-\mathbf{k'})\mathbf{r}}\, d\mathbf{r}. \quad (39.4)$$

The integral in (39.4) is taken over a one unit cell.

Substituting (39.2) into the transport equation, we obtain, in the usual way, the electron mean free path, which is similar to the expression (43.15) in[3],

$$l_{\mathbf{D}} = \frac{2\pi}{a^2 k^2}\left(\frac{dE}{dk}\right)^2 \frac{1}{\displaystyle\int_0^\pi \left|V_n''^{\,\mathbf{kk'}}\right|^2 (1-\cos\vartheta)\sin\vartheta\, d\vartheta}. \quad (39.5)$$

To estimate the dependence of $V_n''^{\,\mathbf{kk'}}$ on k and ϑ, we must remember that the main contribution to the integral (39.4) is made by the constant term of the product $V'' U_{\mathbf{k'}}^* U_{\mathbf{k}}$, equal to the average of this product for a unit cell: $\overline{V'' U_{\mathbf{k'}}^* U_{\mathbf{k}}}$. Taking this quantity outside the integral sign, denoting for brevity $\mathbf{k}-\mathbf{k'}$ by q, and integrating the exponential function over a unit cell in the form of a cube of edge a, we obtain

$$V_n''^{\,\mathbf{kk'}} = \overline{V_n' U_{\mathbf{k'}}^* U_{\mathbf{k}}} \prod_{\alpha=1}^{3} \frac{\sin\dfrac{a q_\alpha}{2}}{\dfrac{a q_\alpha}{2}} \simeq a^3 \overline{V'' U_{\mathbf{k'}}^* U_{\mathbf{k}}}\left(1 - \frac{a^2 q^2}{24}\right) =$$

$$= a^3 \overline{V_n'' U_{\mathbf{k'}}^* U_{\mathbf{k}}}\left[1 - \frac{a^2 k^2}{12}(1-\cos\vartheta)\right]; \quad (39.6)$$

here, q_α is the component of $\mathbf{q}$ along each cube edge. In the above transformation, we have assumed that by virtue of the law of conservation of energy k = k' and $q^2 = 4k^2 \sin^2\dfrac{\vartheta}{2} = 2k^2(1-\cos\vartheta)$.

From the normalization of the functions $U_{\mathbf{k}}$ we find

$$a^3 \overline{V_n' U_{\mathbf{k'}}^* U_{\mathbf{k}}} \simeq \overline{V_n''}, \quad (39.7)$$

where V_n'' is the average value of the perturbation potential in a cell.

Substituting (39.6) into (39.5), integrating with respect to ϑ, and omitting the powers of ak higher than the second, we obtain the mean free path

$$l_{\mathbf{D}} = \frac{\pi}{a^3 k^2}\left(\frac{dE}{dk}\right)^2 \frac{1}{\left|\bar{V}_n''\right|^2 \left(1 - \dfrac{2}{9} a^2 k^2\right)}. \quad (39.8)$$

The lower bar over V_n^* denotes averaging over the n-th unit cell and the upper bar over all unit cells.

It should be noted that since in the derivation of (39.6) we have ignored powers of ak higher than the second, the expression obtained for l_D is valid particularly for semiconductors, in which the quantity ak is small for the great majority of electrons.

§40.   Mechanism of Electron Motion in Liquids

The calculations carried out in the present chapter are based on the assumption that there is an electron mean free path in a liquid. All the experimental data presented in Chap. I indicate that there are no basic differences between electronic conduction in crystals and liquids. Consequently, in liquids, as in crystals, we can use the concept of the electron mean free path.

Moreover, our calculations confirm a posteriori the validity of the concept of a mean free path. For small values of ε, it follows from the formulas given in § 38 that $l_l \gg$ a, where a is the lattice constant. Furthermore, at room and higher temperatures, the "liquid" scattering is less important than the scattering on phonons.

Nevertheless, there is a different point of view held, for example, by Fisher [6]. He rejects the concept of an electron mean path and discusses the motion of an electron in a liquid entirely within the classical theory, assuming that a series of jumps from one atom to another occurs with a probability equal to the quantum-mechanical probability of a transition. This process is assumed to be Markovian, i.e., it is assumed that successive jumps are independent of one another, as in the case of total indeterminacy of the electron momentum. This assumption is not deduced from quantum mechanics but from the absence of local structure repetitions in rectilinear translations. However, in the ξ coordinate system, such a repetition does occur approximately; moreover, if the electrons are tightly bound to the atoms, the repetition follows from the identity of the atoms, irrespective of their position, because the influence of the nearest neighbors is only a small perturbation. Thus, the independence of successive jumps, which is used as the basis of the theory, cannot be regarded as justified.

Moreover, in the jump theory, it is assumed that at any particular moment, an electron is localized near one atom. Such a localization should not occur in the presence of a large number of identical atoms, since it would give

rise to exceptionally large kinetic energy. The steady states of an electron in
a system of interacting identical or almost identical atoms is described by
wave functions which are propagated more or less over the whole system. Such
a collective state is not due to ideal order in the distribution of the atoms but
due to their interactions, and occurs — as proved by the quantum-chemical
calculations for a hydrogen molecule — even in the case of two atoms, when order or disorder in the distribution of the atoms has no meaning. Obviously, in
a disordered system of atoms, the steady-state functions of an electron are not
necessarily plane waves, Bloch or quasi-Bloch waves, but some other waves
propagated over the whole system. Thus, even if an electron is initially localized near one atom, the interaction with neighboring atoms does not lead
to a jump but to a spreading of the electron wave function over many atoms
since any constant perturbations of the initial quantum system make it approach the steady state of the perturbed system.

Strong fluctuations of the short-range order in a liquid may give rise to
local stationary (for bound atoms) electron states, but we shall see (Chap. IX)
that the number of these states is small and that the electron jumps between
these states make a negligible contribution to conduction, compared with the
usual energy-band conduction.

It follows that electron conduction in normal liquids cannot be described
by the jump mechanism. This mechanism may be useful in investigations of
ionic conduction in liquids and particularly in glasses, or of electron conduction due to impurity atoms interacting very weakly with one another.

What is the nature of the motion of an electron in a liquid? We can
answer this question only in general terms.

The method, developed in Chaps. IV and V, of dealing with an electron
in an amorphous substance, using a deformed system of coordinates ξ, makes
it possible to prove the existence of bands in the energy spectrum of the electrons and to determine band edges. However, this method does not give the
band structure and details of the nature of motion of an electron, because it
is not possible to solve the determinant of the system of equations (31.16) and
to find explicitly all the linear combinations of (30.11). Although the initial
function $\psi_\varkappa$ of Eq. (35.1) is formally a Bloch function in the ξ coordinates, it
is not a modulated plane wave. The function $\psi_\varkappa$ describes the motion of an
electron along a quasicrystallographic direction, i.e., along some curve which
makes definite angles with the curvilinear coordinates ξ. We can show that
inertial forces impede the motion of an electron along a curvilinear trajectory
and, therefore, $\psi_\varkappa$ is not a stationary solution with a definite energy. Moreover, the functions $\psi_\varkappa$ do not satisfy the empty-lattice test. We shall assume

that the potential distribution in space is constant for an amorphous substance, but that the depth of the potential minima and maxima tends to zero. The functions $u_\varkappa(\xi)$ tend to a constant value and the wave function should degenerate into a plane wave, while even if $u_\varkappa(\xi) = $ const., $\psi_\varkappa$ is not a plane wave but represents the motion of an electron along a curvilinear trajectory.

It follows that the functions $\psi_\varkappa$ are close to the wave functions of the steady state if the potential wells are sufficiently deep and there is strong local anisotropy, i.e., if an amorphous substance is obtained from a strongly anisotropic crystal. In this case, the electron energy varies considerably with the quasicrystallographic directions, so that inertial forces disturb the motion of an electron less along these directions. An example of such an amorphous substance is a compound consisting of polymer chains. If an electron moves easily from atom to atom in a chain and transitions between loosely-coupled chains are difficult, one of the steady states of an electron is a wave propagating along a chain. This wave is described by the function $\psi_\varkappa$, if one of the ξ axes is directed along the chain. Obviously, this state corresponds to the lowest energy of the valence electrons.

In the present chapter, the electron mean free path in liquids has been calculated by means of different first-approximation wave functions Ψ_k [Eq. (35.2)] which are known as quasi-Bloch functions and satisfy equations which differ by small terms (of the order of ε) from Schrödinger's equation. In the one-dimensional case, $U_k(\xi)$ are periodic functions which have the same properties as $u_k(\xi)$. In the three-dimensional case, $U_k(\xi)$ are no longer strictly periodic functions of ξ since in various cells the vector $\mathbf{k}$ makes different angles with the quasicrystallographic directions. The functions $U_k(\xi)$ reproduce the quasiperiodic variations of the potential in the Cartesian coordinates, thereby allowing for the different directions of the local ξ axes of the potential field with respect to the vector $\mathbf{k}$. The more isotropic the cell the more exactly are the values repeated in different cells.

The functions Ψ_k are modulated plane waves which satisfy the empty-lattice test. Although Ψ_k may be used as the initial state in calculations using the quantum-transition method, they do not represent stationary states even in the zeroth-order approximation. The stationary functions of the zeroth-order approximation should be linear combinations of the functions Ψ_k, similar to linear combinations of (30.11). An important deficiency of the functions Ψ_k is the fact that they are not eigenfunctions of a single operator but belong to different operators, the latter differing from the Schrödinger operator by small terms.

Obviously, the functions Ψ_k best describe stationary states for shallow potential wells and almost-isotropic unit cells; they are quite inapplicable to potentials with deep wells and strong local anisotropy, as in the case of polymers.

In an amorphous substance with small departures from short-range order in the zeroth-order approximation, an electron executes a progressive (not necessarily rectilinear) motion and is represented by a traveling wave modulated in accordance with the potential. These general properties do not depend on the method of propagation and follow from obvious physical considerations.

Therefore, a wave function describing the stationary state of an electron in a liquid may be written in the form:

$$\Phi_k = \frac{1}{\sqrt{G}} u_k(\xi) e^{i \sum_{\alpha=1}^{3} k_\alpha \lambda_\alpha} ; \tag{40.1}$$

here, λ is the electron coordinate in a particular coordinate system, the exponential factor represents the progressive motion of an electron, while the factor $u_k(\xi)$, which is periodic in ξ, describes the influence of the shape of the potential. The coordinates λ are found exactly from the condition that the correct wave functions determined in any way are written in the form of Eq. (40.1). However, since a rigorous calculation of Φ_k is, in practice, impossible, we shall indicate some properties of λ which follow from self-evident physical considerations.

The functions Φ_k describe the propagation of a wave along curves making definite angles with the coordinate lines λ, and they should be intermediate between the functions $\psi_\varkappa$ and Ψ_k. It means that the coordinate lines λ are intermediate between Cartesian axes and the coordinate lines ξ. The directions of the λ lines represent the balance of inertial and potential forces acting on an electron, and depend not only on the geometry of the potential function (like the coordinate lines ξ) but also on the depth of the potential wells.

It follows that in the case of deep potential wells with strong local anisotropy, the coordinates λ approach the coordinates ξ and the functions Φ_k approach the functions $\psi_\varkappa$; conversely, if the potential wells are shallow and the unit cells isotropic, the coordinates λ approach the Cartesian system and the functions Φ_k approach the function Ψ_k. Thus, the functions Φ_k satisfy the empty-lattice test.

We shall formulate these general ideas in a simple quantitative way. For this purpose, we shall assume that the directions of the tangents to the electron trajectory, represented by definite values of k_α, lie within a fixed solid angle ω_1 and also within another solid angle ω_2, related to the ξ coordinate system. Obviously, $\omega_1 + \omega_2 = 4\pi$. The quantity ω_1 or ω_2 represents the approach of the coordinates λ to the coordinates x or ξ, i.e., if $\omega_1 \to 0$, $\omega_2 \to 4\pi$, then $\lambda \to x$, but if $\omega_1 \to 4\pi$, $\omega_2 \to 0$, then $\lambda \to \xi$. For simplicity, we shall assume that the probability of the trajectory directions is distributed uniformly within the solid angles ω_1 and ω_2.

We assume approximately that the solid angle ω_1 lies within a conical surface of angle ϑ_1 the axis of which is a straight line; the trajectory λ approaches this straight line when $\omega_1 \to 0$, $\vartheta_1 \to 0$. Similarly, the angle ω_2 lies within a conical surface of angle ϑ_2, but the axis of this surface has different directions for different cells, coinciding with definite crystallographic directions which the trajectory λ approaches when $\omega_2 \to 0$. The densities of the directions of the axes of both cones should be equal to the square root of the usual direction densities, so that the density of the λ directions, equal to the product of the densities of the directions of the axes of both cones, has the normal value.

The scale of the λ coordinates is also intermediate between the Cartesian and the ξ coordinates. Thus, if, for example, $d\xi = \dfrac{dx}{1 + \varepsilon\gamma}$, as in Chap IV, where γ is a random function of coordinates, then

$$ds_\lambda = \frac{ds}{1 + \varepsilon\eta\gamma} ; \tag{40.2}$$

here, ds is an element of the trajectory λ on the usual scale, ds_λ is the same element in the λ coordinates, and η is a proper fraction.

Thus, the quantities η and ω_1 or ϑ_1 represent the deviations of the λ coordinates from a Cartesian system.

In conclusion, we note that the wave function of an electron in an amorphous substance may be written in the form of Eq. (40.1), even when departures from the short-range order are large so that the quasicrystallographic directions disappear and the fluctuations in the distances between neighboring atoms are large. In this case, the direction of the trajectory with a given k_α near an atom is determined by the random position of its neighbors. Thus, for example, the trajectory corresponding to the lower edge of an allowed band should pass through each atom and its two nearest neighbors. For large fluctuations in distances between neighboring atoms, this will obviously give considerable broadening of the allowed bands which may fill the forbidden bands. The gradual transition from quasicrystalline order to complete disorder, which occurs in a liquid on heating from the melting point to the critical temperature, needs special discussion. This problem will be dealt with for liquid metals in Chap. X.

Chapter VII

SCATTERING OF ELECTRONS

ON THERMAL VIBRATIONS IN A LIQUID [1]

§41. Initial Equations

In the preceding chapter, we dealt with the scattering of electrons on the departures from long-range order in liquids. However, apart from this specifically "liquid" scattering, there is, as in crystals, the scattering of electrons on thermal vibrations, which may predominate if the melting point is high. In the present chapter, we shall consider this type of scattering.

In Chap. VI, we showed that a satisfactory zeroth-order approximation for the description of electrons in a liquid is given by quasi-Bloch wave functions of the type:

$$\Psi_k = \frac{U_k(\xi)}{\sqrt{G}}\, e^{i\mathbf{k}\cdot\mathbf{r}}, \tag{41.1}$$

where $\mathbf{k}$ is the electron wave vector and $U_k(\xi)$ are functions which are almost periodic in a deformed coordinate system ξ, in which the self-consistent potential for an electron in a liquid is periodic.

In the theory of electrical conduction, thermal vibrations are considered in the elastic continuum approximation and the discrete structure of the lattice is taken into account only in determining the limiting wave number of phonons. In this approximation, there is no difference between a crystal and a liquid, so that we can use as phonon wave functions the same functions as for solids.

Knowing the unperturbed wave functions (41.1) for the electrons, and employing the usual functions for the phonons, we can repeat for liquids the whole calculation given for solid conductors in § 15.

If we assume that the interaction between the electrons and phonons is a small perturbation, the eigenfunction for the unperturbed problem is a product (15.1) of an electron wave function (41.1) and the functions of all the

156

lattice oscillators; the wave function for the perturbed case may be expanded in a series (15.3) of these products:

$$\Psi(t) = \sum c(\mathbf{k},\ N_{\mathbf{q}j},\ t)\,\psi_{\mathbf{k}}(\mathbf{r}) \times$$

$$\times \prod_{\mathbf{q}} \prod_{j} \psi_{N_{\mathbf{q}j}} e^{-\frac{it}{\hbar}\left[E_{\mathbf{k}} + \sum_{\mathbf{q}} \sum_{j} \left(N_{\mathbf{q}j} + \frac{1}{2} \right) \hbar\omega_{\mathbf{q}j} \right]}. \tag{41.2}$$

According to Eq. (15.4), the expansion coefficients c depend on time and for small values of t obey the equation

$$i\hbar\dot{c}(\mathbf{k}',\ N_{\mathbf{q}j}) = \int \psi^*(\mathbf{k}',\ N_{\mathbf{q}j})\, U\psi'(\mathbf{k},\ N_{\mathbf{q}j})\, d\tau \times$$

$$\times\, e^{\frac{it}{\hbar}\left[E_{\mathbf{k}'} - E_{\mathbf{k}} + \sum_{\mathbf{q}} \sum_{j} \left(N'_{\mathbf{q}j} - N_{\mathbf{q}j} \right) \hbar\omega_{\mathbf{q}j} \right]}. \tag{41.3}$$

The usual notation is employed in the above equation; $\mathbf{k}$ and $\mathbf{k}'$ are the electron wave vectors; $\mathbf{q}$ is the phonon wave vector; j is the direction of the phonon polarization; $N_{\mathbf{q}j}$ is the number of phonons with the wave vector $\mathbf{q}$ and the polarization direction j; $E_{\mathbf{k}}$ and $E_{\mathbf{k}'}$ are the energies of electrons for the corresponding values of the wave vector; $\omega_{\mathbf{q}j}$ is the frequency of a phonon with given values of $\mathbf{q}$ and j. Integration with respect to $d\tau$ includes integration with respect to the electron radius vector $\mathbf{r}$, and with respect to the oscillator coordinates $a_{\mathbf{q}j}$.

The perturbation energy U is different for acoustical and optical thermal vibration modes.

We shall consider first the scattering on acoustical vibrations: they are the only vibrations present in metals and covalent semiconductors.

§ 42. Change in the Electron Distribution Function Under the Action of Acoustical-Mode Vibrations

In the deformable ion approximation, the perturbation energy is

$$U = -\mathbf{u} \cdot \nabla V, \tag{42.1}$$

where V is the self-consistent potential for an electron, and $\mathbf{u}$ is the displacement of ions or atoms in thermal vibration

$$\mathbf{u} = \frac{1}{\sqrt{G}} \sum_{\mathbf{q}} \sum_{j=1}^{3} \mathbf{e}_{\mathbf{q}j} \left(a_{\mathbf{q}j} e^{i\mathbf{q} \cdot \mathbf{r}} + a_{\mathbf{q}j}^* e^{-i\mathbf{q} \cdot \mathbf{r}} \right); \qquad (42.2)$$

G is the number of atoms in the region being considered; and $\mathbf{e}_{\mathbf{q}j}$ is a unit vector representing the polarization of the thermal vibrations. After substituting Eqs. (42.1) and (42.2), the integral in Eq. (41.3) becomes [2]:

$$J = \frac{1}{\sqrt{G}} \sum_{\mathbf{q}} \sum_{j=1}^{3} (J_{\mathbf{q}j}^+ + J_{\mathbf{q}j}^-), \qquad (42.3)$$

where

$$J_{\mathbf{q}j}^+ = \mathbf{e}_{\mathbf{q}j} \int \Psi_{\mathbf{k}'}^* (\mathbf{r}) \nabla V e^{i\mathbf{q} \cdot \mathbf{r}} \Psi_{\mathbf{k}} (\mathbf{r}) \, d\mathbf{r} \times$$

$$\times \int \psi_{N_{\mathbf{q}j}}' (a_{\mathbf{q}j}') a_{\mathbf{q}j} \psi_{N_{\mathbf{q}j}} \, da_{\mathbf{q}j} \prod_{\mathbf{p}, \, k \ne \mathbf{q}, \, j} \int \psi_{N_{\mathbf{p}k}}' (a_{\mathbf{p}k}) \psi_{N_{\mathbf{p}k}} (a_{\mathbf{p}k}) \, da_{\mathbf{p}k}. \quad (42.4)$$

A similar expression is obtained for $J_{\mathbf{q}j}^-$. The integrals with respect to the oscillator coordinates $a_{\mathbf{q}j}$ and $a_{\mathbf{p}k}$ are not equal to zero only if

$$N_{\mathbf{p}k}' = N_{\mathbf{p}k}, \quad N_{\mathbf{q}j}' = N_{\mathbf{q}j} - 1 \quad \text{for} \quad J_{\mathbf{q}j}^+, \quad N_{\mathbf{q}j}' = N_{\mathbf{q}j} + 1 \quad \text{for} \quad J_{\mathbf{q}j}^-,$$

and integration with respect to $a_{\mathbf{q}j}$ gives the factors

$$\sqrt{\frac{\hbar}{2M\omega_{\mathbf{q}j}} N_{\mathbf{q}j}} \quad \text{for} \quad J_{\mathbf{q}j}^+, \text{and} \quad \sqrt{\frac{\hbar}{2M\omega_{\mathbf{q}j}} (N_{\mathbf{q}j} + 1)} \quad \text{for} \quad J_{\mathbf{q}j}^-,$$

where M is the mass of an atom.

The integral with respect to the electron coordinates in Eq. (41.3) may, after the substitution of the expression (41.1) for $\Psi_{\mathbf{k}} (\mathbf{r})$ and $\Psi_{\mathbf{k}'} (\mathbf{r})$, be represented in the form of a sum of integrals over G deformed unit cells:

$$\mathbf{K}^\pm = \frac{1}{G} \sum_{n} e^{i \, (\mathbf{k} \pm \mathbf{q} - \mathbf{k}') \cdot \mathbf{R}_n} \times$$

$$\times \int_1 e^{i \, (\mathbf{k} \pm \mathbf{q} - \mathbf{k}') \cdot \mathbf{r}} \nabla V (\mathbf{r}') U_{\mathbf{k}} (\mathbf{r}') U_{\mathbf{k}'}^* (\mathbf{r}') \, d\mathbf{r}, \qquad (42.5)$$

where $\mathbf{R}_n$ is the radius vector of an n-th site.

For a crystal, the integral over a unit cell is independent of the "serial" number of this cell, because of the periodicity of $V(\mathbf{r})$ and $U_k(\mathbf{r})$, and because the values of the vector $\mathbf{R}_n$ form a regular lattice in space. Therefore, the sum of n terms is different from zero only if

$$\mathbf{k} \pm \mathbf{q} = \mathbf{k}', \tag{42.6}$$

where the "+" sign refers to K^+ and the "−" sign to K^-. The physical condition (42.6) represents the law of conservation of momentum in the electron–phonon system, because $J_{\mathbf{q}j}^+$ represents phonon absorption and $J_{\mathbf{q}j}^-$ represents phonon emission. If the condition (42.6) is satisfied, summation up to n simply yields the factor G.

As mentioned in Chap. VI, in the isotropic approximation $U_\mathbf{k}$ for a liquid is a periodic function of the deformed coordinates ξ, and if the condition (42.6) is obeyed, the integral for a deformed unit cell is independent of the "serial" number of this cell to within small quantities of the order of ε, which is the degree of departure from short-range order, i.e., the relative deformation of the cell. There is no point in introducing corrections of the order of ε since the perturbation itself is small and these corrections would be of the second order of smallness.

Thus, if the condition (42.6) is satisfied, the same result is obtained for a liquid as for a crystal.

All further calculations given for crystals in § 15 will now be valid for a liquid, to within corrections of the order of ε, and will give the same results. We are dealing here purely with the phonon scattering of electrons, which is completely independent of the departures from long-range order in a liquid.

However, if the condition (42.6) is not satisfied, then in the case of a liquid we cannot say that the sum of n terms in Eq. (42.5) will be equal to zero. First, in a liquid, unit cells which are far apart are rotated with respect to one another by arbitrary angles and, therefore, the corresponding vectors $\mathbf{r}'$ make different angles with the constant vector $\mathbf{k} \pm \mathbf{q} - \mathbf{k}'$, i.e., the exponential factors in the integrals and, therefore, the integrals themselves depend on the "serial" number of the cell. Secondly, the vectors $\mathbf{R}_n$ for a liquid do not form a regular lattice but are random quantities.

Thus, if the condition (42.6) is not obeyed, the summation in Eq. (42.5) may be carried out in accordance with the rules for random quantities and the sum will be different from zero. We shall show that the summation over G deformed electron cells gives the factor $G\varepsilon^2$. In the isotropic approximation, the integral for a unit cell in Eq. (42.5) can be taken outside the summation symbol and the problem reduces to a calculation of the sum

$$\Sigma = \sum_n e^{i\mathbf{q}' \cdot \mathbf{R}_n}, \qquad (42.7)$$

where, for brevity, it is assumed that $\mathbf{k} \pm \mathbf{q} - \mathbf{k}' = \mathbf{q}'$.

The vectors $\mathbf{R}_n$ for a liquid are random quantities but they are completely independent of one another, so that the moduli of the differences of the vectors $\mathbf{R}_n$ for neighboring cells differ from the lattice constant by small corrections of the order of ε.

We shall first calculate such a sum for the one-dimensional model of a liquid:

$$\sum_1 = \sum_{n=1}^{G} e^{qx_n}, \qquad (42.8)$$

where

$$qa = \frac{2k\pi}{G}, \quad x_n = x_{n-1} + a(1 + \varepsilon\gamma_n), \qquad (42.9)$$

a is the lattice constant, k = 0, 1, 2, ..., G is the number of atoms in a chain, $\varepsilon \ll 1$, and γ_n are random quantities. The calculation of the sum is very similar to the calculation of the length of a molecule with links whose rotation is restricted (see, for example, pp. 409-411 in [3]) and, therefore, we shall use similar steps in the calculation. We shall calculate the square of the modulus of the sum:

$$\left|\sum_1\right|^2 = \sum_{n=1}^{G} e^{iqx_n} \sum_{m=1}^{G} e^{-iqx_m} = G + \sum_{n<m} \left[e^{iq(x_n - x_m)} + e^{iq(x_m - x_n)}\right].$$
$$(42.10)$$

From Eq. (42.9), it follows that

$$x_n = na + a\varepsilon \sum_{l=1}^{n} \gamma_l, \quad x_m = ma + a\varepsilon \sum_{l=1}^{m} \gamma_l. \qquad (42.11)$$

Substituting Eq. (42.11) into (42.10), we shall rewrite the latter in the form:

$$\left|\sum_1\right|^2 = G + \sum_{n<m} \left\{ e^{iqa(n-m)} \left[\cos\left(\varepsilon aq \sum_{l=n+1}^{m} \gamma_l\right) - i \sin\left(\varepsilon aq \sum_{l=n+1}^{m} \gamma_l\right) \right] + \right.$$

$$+ e^{iqa(m-n)} \left[\cos \left(\varepsilon a q \sum_{l=n+1}^{m} \gamma_l \right) + i \sin \left(\varepsilon a q \sum_{l=n+1}^{m} \gamma_l \right) \right] \bigg\} .$$

$$\text{(42.12)}$$

Changing now to average values, we find that since γ_l is equally likely to be positive or negative, sines become zero and the average cosines in the sum become equal to the products of the average cosines of the terms. Since the rms values of all γ_l are the same, the product reduces to a power of the quantity $\varepsilon a q \gamma_1$. Consequently

$$\overline{\left| \sum_{1} \right|^2} = G + \sum_{n<m} \overline{\left[\cos (\varepsilon a q \gamma_l) \right]}^{m-n} \left[e^{iqa(m-n)} + e^{iqa(n-m)} \right]. \quad \text{(42.13)}$$

The summation is easy and gives

$$\overline{\left| \sum_{1} \right|^2} = G + \frac{G \zeta \eta}{1 - \zeta \eta} + \frac{G \dfrac{\zeta}{\eta}}{1 - \dfrac{\zeta}{\eta}} - \frac{\zeta \eta (1 - \zeta^G \eta^G)}{(1 - \zeta \eta)^2} - \frac{\zeta}{\eta} \frac{1 - \dfrac{\zeta^G}{\eta^G}}{\left(1 - \dfrac{\zeta}{\eta} \right)^2},$$

$$\text{(42.14)}$$

where, for brevity, we use

$$\eta = e^{iqa} = e^{2\pi i \frac{k}{G}}, \quad \zeta = \overline{\cos (\varepsilon a q \gamma_l)}. \quad \text{(42.15)}$$

Dropping the last two terms in Eq. (42.14), which do not contain the large factor G, we obtain

$$\overline{\left| \sum_{1} \right|^2} \simeq G \frac{1 - \zeta^2}{1 - \zeta \eta - \dfrac{\zeta}{\eta} + \zeta^2} . \quad \text{(42.16)}$$

Expanding in series the expressions (42.15) and bearing in mind that ε is small and that $\overline{\gamma_l^2} = 1$, we find that from Eq. (42.16) for $q \neq 0$:

$$\overline{\left| \sum_{1} \right|^2} \simeq G \varepsilon^2. \quad \text{(42.17)}$$

If $q = 0$, the approximate expression (42.16) should not be used and the more exact expression (42.14) gives $\overline{\left| \sum_{1} \right|^2} = G^2$ in accordance with the usual theory.

The three-dimensional case can be reduced to the one-dimensional treatment by directing the x axis along the vector q' and dividing the region containing G atoms into $G^{2/3}$ chains of atoms parallel to the x axis, with $G^{1/3}$ atoms in each chain. Then, summation along a chain can be carried out as in the one-dimensional case, and this gives the factor $G^{1/3} \varepsilon^2$ in the expression for the average of the modulus squared. The summation for $G^{2/3}$ chains gives the factor $G^{2/3}$; consequently, the summation reduces to multiplication by $G\varepsilon^2$.

Thus, electrons may be scattered by phonons in a liquid without obeying the law of conservation of momentum within the electron—phonon system. This is because the lattice, lacking some long-range order, is capable of absorbing additional momentum when a phonon is emitted or absorbed by an electron. A similar effect occurs in the case of impurities in crystals.

We shall call this additional scattering of the electrons the "phonon—liquid" scattering since, in this case, thermal vibrations and departures from long-range order are equally important. We shall calculate the mean free path $l_{ph. l}$ due to this phonon—liquid scattering. At first sight, it seems inconsistent to allow for the phonon—liquid scattering since in the case of the normal phonon scattering in a liquid we have neglected corrections of the order of ε. However, these were only small corrections to the numerical values of coefficients, while the phonon—liquid scattering is a new physical effect with characteristic features and a different dependence on the electron energy; therefore, this effect may, in some cases, be important in spite of the smallness of ε.

Due to the exponential factor under the integral sign for a unit cell, this integral depends, strictly speaking, on $\mathbf{k} + \mathbf{q} - \mathbf{k}'$. However, the factor ∇V is maximal within the atomic core where $\mathbf{r}'$ is small and, therefore, the influence of the exponential factor is small (see p. 227 in [2]) and we shall ignore it. The inclusion of this exponential factor would give a coefficient close to unity.

Integration with respect to oscillator coordinates is carried out as in the theory of metals. Integrating (41.3) with respect to time, we obtain, after some transformations(§ 15),

$$|c(\mathbf{k}', N_{qj} - 1, t)|^2 = \frac{2C^2 \varepsilon^2}{9G^2 M\hbar} (\mathbf{k} - \mathbf{k}') \times$$
$$\times \mathbf{e}_{qj} N_{qj} \Omega (E_{\mathbf{k}'} - E_{\mathbf{k}} - \hbar\omega_{qj}); \qquad (42.18)$$

$$|c(\mathbf{k'}, N_{\mathbf{q}j}+1, t)|^2 = \frac{2C^2\varepsilon^2}{9G^2M\hbar}(\mathbf{k}-\mathbf{k'}) \times$$

$$\times \mathbf{e}_{\mathbf{q}j}(N_{\mathbf{q}j}+1)\,\Omega\,(E_{\mathbf{k'}}-E_{\mathbf{k}}+\hbar\omega_{\mathbf{q}j}); \qquad (42.19)$$

here,

$$C = \frac{\hbar^2}{2m}\int_1 |\nabla U|^2\,d\mathbf{r}; \quad \Omega(x) = 2\,\frac{1-\cos\dfrac{xt}{\hbar}}{\left(\dfrac{x}{\hbar}\right)^2}. \qquad (42.20)$$

Since, in contrast to the normal scattering, $\mathbf{k}-\mathbf{k'} \neq \mp \mathbf{q}$, electrons interact not only with longitudinal but also with transverse waves. Since the function $\Omega(x)$ is of the δ type, the law of conservation of energy in the electron—phonon system is obeyed. This means that the disturbed lattice absorbs additional momentum but does not absorb additional phonon energy.

In order to find the change in the electron distribution function due to scattering, it is necessary to multiply Eqs. (42.18) and (42.19) by the probability of achieving the initial state and the probability of the final state being vacant, and then to sum over all the initial values of the electron wave number $\mathbf{k}$, over all values of the phonon wave number $\mathbf{q}$, and over the directions of the phonon polarization. In contrast to the usual phonon scattering, the summation for all $\mathbf{k}$ should be carried out independently of the summation for all $\mathbf{q}$. Consequently, we obtain

$$f(\mathbf{k'}, t)-f(\mathbf{k'}, 0) = \frac{2C^2\varepsilon^2}{9G^2M\hbar}\sum_{\mathbf{k}}\sum_{\mathbf{q}}\sum_{j}[(\mathbf{k}-\mathbf{k'})\cdot\mathbf{e}_{\mathbf{q}j}]^2 \times$$

$$\times \{\Omega\,(E_{\mathbf{k'}}-E_{\mathbf{k}}+\hbar\omega_{\mathbf{q}j})[f(\mathbf{k})(1-f(\mathbf{k'}))[(N_{\mathbf{q}j}+1)-$$

$$-f(\mathbf{k'})(1-f(\mathbf{k}))N_{\mathbf{q}j}]+\Omega\,(E_{\mathbf{k'}}-E_{\mathbf{k}}-\hbar\omega_{\mathbf{q}j}) \times$$

$$\times [f(\mathbf{k})(1-f(\mathbf{k'}))N_{\mathbf{q}j}-f(\mathbf{k'})(1-f(\mathbf{k}))(N_{\mathbf{q}j}+1)]\}.$$

$$(42.21)$$

§43. Bloch's Integral Equation

We shall introduce in the **k** space the spherical system of coordinates k, θ, Φ with its axis along k_X, and in the **q** space the system q, ϑ, φ with its axis along **k** — **k′**. We shall replace summation over **k** and **q** by integration using the formulas

$$\sum_{\mathbf{k}} F(\mathbf{k}) = \mathfrak{Q}_0 \frac{G}{(2\pi)^3} \int k^2 dk \sin\theta d\theta d\Phi F(\mathbf{k}); \left.\begin{array}{c}\\\\\\\end{array}\right\}$$

$$\sum_{\mathbf{q}} F(\mathbf{q}) = \mathfrak{Q}_0 \frac{G}{(2\pi)^3} \int q^2 dq \sin\vartheta d\vartheta d\varphi F(\mathbf{q}); \tag{43.1}$$

Ω_0 is the volume of a unit cell.

We shall assume that the phonons are in thermal equilibrium, and that the electrons depart a little from thermal equilibrium due to an electric field or a temperature gradient acting along the x axis. In this case,

$$N_{\mathbf{q}j} = \frac{1}{e^{\frac{\hbar\omega_{\mathbf{q}j}}{k_B T}} - 1}; \quad f = f_0 + k_x f_1(E); \quad f_0 = \frac{1}{e^{\frac{E-\mu}{k_B T}} + 1}; \tag{43.2}$$

$f_1(E)$ is a small correction the square of which we shall neglect; and μ is the Fermi potential. The energy will be regarded as depending only on the modulus of **k** or **k′**.

Integration with respect to φ gives the factor 2π; ϑ occurs only in the scalar product **(k** — **k′)** · $\mathbf{e}_{\mathbf{q}j}$. Let us assume that j = 1 represents a longitudinal wave and j = 2, 3 represent transverse waves, where $\mathbf{e}_{\mathbf{q}2}$ lies in the **(k** — **k′)** · **q** plane, and $\mathbf{e}_{\mathbf{q}3}$ is at right-angles to it. Then, **(k** — **k′)** · $\mathbf{e}_{\mathbf{q}_1}$ = | **k** — **k′** | cos ϑ, **(k** — **k′)** · $\mathbf{e}_{\mathbf{q}2}$ = | **k** — **k′** | sin ϑ, and **(k** — **k′)** · $\mathbf{e}_{\mathbf{q}3}$ = 0; and also

$$\int [(\mathbf{k} - \mathbf{k'}) \cdot \mathbf{e}_{\mathbf{q}_1}]^2 \sin\vartheta d\vartheta = \frac{2}{3}(\mathbf{k} - \mathbf{k'})^2, \left.\begin{array}{c}\\\\\end{array}\right\}$$

$$\int [(\mathbf{k} - \mathbf{k'}) \cdot \mathbf{e}_{\mathbf{q}2}]^2 \sin\vartheta d\vartheta = \frac{4}{3}(\mathbf{k} - \mathbf{k'})^2. \tag{43.3}$$

Integrating with respect to q, we shall take into account the δ-nature of the function Ω(x). All the factors, apart from Ω, may be placed outside the symbol of integration with respect to q, substituting in these factors

$$\omega_{qj} = \pm \frac{1}{\hbar}(E_k - E_{k'}), \quad q = \frac{\omega_{qj}}{w_j} = \pm \frac{1}{\hbar w_j}(E_k - E_{k'}), \quad (43.4)$$

where w_1 and w_2 are, respectively, the velocities of longitudinal and transverse waves. The resultant integral gives

$$\int_0^\infty \mathcal{Q}(E_{k'} - E_k - \hbar\omega_{qj}) dq \simeq \frac{2\pi t}{w_j}. \quad (43.5)$$

When Eqs. (43.2) and (43.4) are substituted into (42.21), all the terms of the zeroth order of smallness are canceled, leaving only the terms with $f_1(E_k)$ and $f_1(E_{k'})$.

It is then evident that

$$(\mathbf{k} - \mathbf{k'})^2 = k'^2 + k^2 - 2kk' \cos\vartheta', \quad (43.6)$$

where ϑ' is the angle between $\mathbf{k}$ and $\mathbf{k'}$. If we take the angle φ from the plane $\mathbf{k'}x$,

$$\cos\vartheta = \cos\theta \cos\theta' + \sin\theta \sin\theta' \sin\varphi. \quad (43.7)$$

In the integration with respect to φ, the term with $\sin\varphi$ disappears and all the other terms are multiplied by 2π.

Substituting into Eq. (42.21) the expressions (43.1)-(43.7), bearing in mind the remarks made about this integration, and differentiating with respect to time, we obtain

$$\left(\frac{\partial f}{\partial t}\right)_{ph.l} = \frac{C^2\mathcal{Q}_0\varepsilon^2}{54\hbar^2\pi^3} \sum_{j=1}^2 \frac{j}{w_j^3} \left\{ \int_{k'}^{k_{max}^j} k^2 dk \int_0^\pi \sin\theta \, d\theta \times \right.$$

$$\times (E_k - E_{k'})(k^2 + k'^2 - 2kk'\cos\theta\cos\theta')\left[k_x f_1(E_k)(1 + N_{qj} - \right.$$

$$\left. - f_0(E_{k'})) - k'_x f_1(E_{k'})(N_{qj} + f_0(E_k))\right] +$$

$$+ \int_{k_{min}^j}^{k'} k^2 dk \int_0^\pi \sin\theta \, d\theta \, (E_{k'} - E_k)(k^2 + k'^2 - 2kk'\cos\theta\cos\theta') \times$$

$$\times \left[k_x f_1(E_k)(N_{qj} + f_0(E_{k'})) - k'_x f_1(E_{k'})(1 + N_{qj} - f_0(E_k))\right]\}.$$

$$(43.8)$$

The limits of integration with respect to k are found from the conditions

$$E(k_{max}^j) = E_{k'} + h\omega_j; \quad E(k_{min}^j) = \begin{cases} E_k - \hbar\omega_j \\ 0 \end{cases} \text{ for } \quad E_k \gtrless \hbar\omega_j;$$

(43.9)

ω_1 and ω_2 are the limiting frequencies of the longitudinal and transverse vibrations. We note that $k'_x = k' \cos\theta'$, $k_x = \cos\theta$.

Carrying out the integration with respect to θ, substituting the values $(\partial f/\partial t)_{ph. \, l}$ into the transport equation for an electric field E_x acting along the x axis, and dividing by $k' \cos\theta' = k'_x$, we obtain Bloch's integral equation for the function $f_1(E)$:

$$\sum_{j=1}^{2} \frac{j}{w_j^2} \int_{k'}^{k_{max}^j} k^2 dk (E_k - E_{k'}) \left\{ \frac{2}{3} k^2 f_1(E_k)[1 + N_{qj} - f_0(E_k)] + \right.$$
$$\left. + (k^2 + k'^2) f_1(E_k)[N_{qj} + f_0(E_k)] \right\} +$$

$$+ \sum_{j=1}^{2} \frac{j}{w_j^3} \int_{k_{min}^j}^{k'} k^2 dk (E_{k'} - E_k) \left\{ \frac{2}{3} k^2 f_1(E_k)[N_{qj} + f_0(E_{k'}) + \right.$$
$$\left. + (k^2 + k'^2) f_1(E_k)[1 + N_{qj} - f_0(E_k)] \right\} =$$

$$= \frac{27\pi^3 \hbar e E_x}{C^2 \Omega_0^2 \epsilon^2} \cdot \frac{\partial f_0}{\partial E} \frac{1}{k'} \cdot \frac{dE}{dk'}.$$

(43.10)

This equation may be simplified by the substitution

$$f_1 = -\chi(E) \frac{df_0}{dE},$$

(43.11)

and after some transformations, which use the explicit expressions for N and f_0, it becomes:

$$\sum_{j=1}^{2} \frac{j}{w_j^3} \left\{ \int_{k'}^{k_{max}^j} N_{qj} k^2 dk (E_k - E_{k'}) e^{\frac{E_k - E_{k'}}{k_B T}} \frac{f_0(E_k)}{f_0(E_{k'})} \times \right.$$

$$\times \left[\frac{2}{3} k^2 \chi(E_F) + (k^2 + k'^2) \chi(E_{k'}) \right] dk + \int_{k_{min}^j}^{k'} N_{qj} k^2 dk \times$$

$$\times (E_{k'} - E_k) \frac{f_0(E_k)}{f_0(E_{k'})} \left[\frac{2}{3} k^2 \chi(E_k) + (k^2 + k'^2) \chi(E_k) \right] \Big\} =$$

$$= \frac{27 M \pi^3 \hbar e E_x}{k' C^2 \Omega_0^2 \cdot \varepsilon^2} \cdot \frac{dE}{dk'} . \tag{43.12}$$

It is easy to extend the above treatment to the case when the action of an electric field is accompanied by the action of a temperature or concentration gradient.

§44. Mean Free Path of an Electron Associated with the Phonon − Liquid Scattering

To calculate the integral in Eq. (43.12), it is convenient to use the notation:

$$\left.\begin{aligned}
\frac{|E_k - E_{k'}|}{k_B T} = y, \quad \frac{E_{k'}}{k_B T} = \eta, \quad \frac{E_{k'} - \mu}{k_B T} = \zeta, \\
\frac{\hbar \omega_j}{k_B T} = y_j, \quad k' = k(\eta), \quad k = k(\eta + y)
\end{aligned}\right\} \tag{44.1}$$

and integrate with respect to y. The left-hand part of Eq. (43.12) then becomes:

$$\frac{(k_B T)^2 k'^4 \chi(E_{k'})}{\dfrac{dE_{k'}}{dk'}} \sum_{j=1}^{2} \frac{j}{w_j^3} (I_j' + I_j''), \tag{44.2}$$

where

$$\left.\begin{aligned}
I_j' &= \int_0^{y_j} \frac{y e^y}{e^y - 1} \cdot \frac{e^\zeta + 1}{e^{\zeta + y} + 1} \left[\frac{2}{3} \cdot \frac{k^2(\eta + y)}{k^2(\eta)} \cdot \frac{\chi(\eta + y)}{\chi(\eta)} + \right. \\
&\quad \left. + \frac{k^2(\eta + y)}{k^2(\eta)} + 1 \right] \frac{dk(\eta + y)}{dk(\eta)} \, dy, \\[2ex]
I_j'' &= \int_0^{y^*} \frac{y}{e^y - 1} \cdot \frac{e^\zeta + 1}{e^{\zeta + y} + 1} \left[\frac{2}{3} \cdot \frac{k^2(\eta + y)}{k^2(\eta)} \cdot \frac{\chi(\eta + y)}{\chi(\eta)} + \right. \\
&\quad \left. + \frac{k^2(\eta + y)}{k^2(\eta)} + 1 \right] \frac{dk(\eta + y)}{dk(\eta)} \, dy,
\end{aligned}\right\} \tag{44.3}$$

and, according to Eq. (43.9), $y^* = y_j$ for $\eta > y_j$, $y^* = \eta$ for $\eta < y_j$. The expressions in front of the square brackets in Eq. (44.3) have their maximum values when $y = 0$ and then they decrease rapidly. Therefore, we shall not be committing a great error if we replace the slowly varying functions $k^2(\eta + y)$, $\chi(\eta + y)$, and $dk(\eta + y)$ with their values for $y = 0$. This is definitely permissible in the case of metals, as well as in the case of semiconductors at high temperatures, because in both instances $\eta \gg y$ for the majority of the electrons and the functions just referred to vary within relatively narrow limits. These assumptions are less valid for semiconductors at low temperatures but in this case the phonon—liquid scattering is not important and a rough estimate only is sufficient.

The assumption $\chi = $ const. is also used in the theory of electrical conduction in crystals at high temperatures. In our case, the structure of Bloch's integral equation is much better suited to this assumption and, therefore, we can extend it to any temperature. This means that the electron mean free path associated with the phonon—liquid scattering exists at any temperature.

On the basis of this assumption, the expressions for I'_j and I''_j simplify greatly:

$$
\left.
\begin{aligned}
I'_j &= \frac{8}{3} \int_0^{y_j} \frac{y e^y}{e^y - 1} \cdot \frac{e^z + 1}{e^{z+y} + 1} \, dy, \\
I''_j &= \frac{8}{3} \int_0^{y^*} \frac{y}{e^y - 1} \cdot \frac{e^z + 1}{e^{z+y} + 1} \, dy.
\end{aligned}
\right\}
\qquad (44.4)
$$

Substituting Eq. (44.2) into Eq. (43.12), solving the latter equation for $\chi(E_{k'})$ and assuming that $\Omega_0 = a^3$, we obtain the electron mean free path governed by the phonon—liquid scattering:

$$
l_{\mathrm{ph.}} \overline{T} = \frac{k' \chi(E_{k'})}{e E_x} = \frac{27 M \pi^3 \hbar}{k'^4 (k_B T)^2 C^2 \varepsilon^2 a^6 \sum_{j=1}^{2} \frac{j}{w_j^3} (I'_j + I''_j)} ; \quad (44.5)
$$

I'_j is a function of y_j, I''_j is also a function of y_j if $\eta > y_j$, if $\eta < y_j$, it is a function of η. According to Eq. (44.1), $y_i = \theta/T$, where θ is the Debye temperature. If we assume that the limiting value of the wave number q_0 is the same for longitudinal and transverse waves, it follows that $y_2 = \dfrac{\theta}{T} \cdot \dfrac{w_2}{w_1}$, and, consequently, I'_j and I''_j are both functions of θ/T. The explicit form of these functions can be obtained for high ($T \gg \theta$) and low ($T \ll \theta$) temperatures.

If $T \gg \theta$, $y \ll 1$, $e^y \simeq 1$, $e^y - 1 \simeq y$, then $y_j < \eta$ for the majority of electrons, i.e., $y^* = y_j$ and

$$I'_1 = I''_1 = \frac{8}{3} \cdot \frac{\theta}{T}, \quad I'_2 = I''_2 = \frac{8}{3} \cdot \frac{\theta}{T} \cdot \frac{w_2}{w_1}. \qquad (44.6)$$

Substituting Eq. (44.6) into Eq. (44.5) and using the relationships

$$w_1 = \frac{k_B \theta}{\hbar q_0}; \quad q_0 = \frac{2\pi}{a} \sqrt[3]{\frac{3}{4\pi}}, \qquad (44.7)$$

we obtain, after making some simple transformations,

$$l_{ph.l} = \frac{9\pi M a k_B \theta}{32 \beta C^2 \varepsilon^2 \hbar^2 (ak)^4} \frac{\theta}{T} \left(\frac{dE}{dk}\right)^2; \qquad (44.8)$$

here, β is a coefficient close to unity:

$$\beta = \frac{1}{3} + \frac{2}{3} \cdot \frac{w_1^2}{w_2^2}. \qquad (44.9)$$

For $T \ll \theta$, we must deal separately with metals and semiconductors. In
the case of metals, only the electrons in a narrow band near the Fermi level
are important and, therefore, we assume that $\zeta = 0$. Next, $\eta < y_j$, $y^* = y_j$ and
since $y_j \gg 1$, the upper limit in both integrals of Eq. (44.4) can be replaced
by infinity. Consequently, we find that

$$I'_j = I''_j = \frac{16}{3} \int_0^\infty \frac{y \, dy}{e^y - e^{-y}} = \frac{16}{3} \cdot \frac{\pi^2}{8} = \frac{2\pi^3}{3} \qquad (44.10)$$

(see pp. 134 and 350 in [4]), I'_j and I''_j are independent of j,

$$l_{ph.l} = \frac{9 M a k_B \theta}{8\pi \beta' C^2 \varepsilon^2 \hbar^2 (ak)^4} \left(\frac{\theta}{T}\right)^2 \left(\frac{dE}{dk}\right)^2, \qquad (44.11)$$

where

$$\beta' = \frac{1}{3} + \frac{2}{3} \cdot \frac{w_1^3}{w_2^3}. \qquad (44.12)$$

In the case of semiconductors $\zeta \gg 1$ and we can neglect unity compared
with e^ζ. In the integral I'_j, we can again replace the upper limit with ∞. As
a result of this, we obtain

$$I'_j = \frac{8}{3} \int\limits_0^\infty \frac{y}{e^y - 1} dy = \frac{8}{3} \cdot \frac{\pi^2}{6} = \frac{4}{9} \pi^2 \qquad (44.13)$$

(see pp. 134 and 244 in [2]).

For the majority of electrons in the conduction band of a semiconductor, η is of the order of 1; whenT $\ll \theta$, $\eta < y_j$, i.e., $y^* = \eta$, and we can easily see that when

$$\eta < \frac{\pi^2}{6}, I''_j \leqslant \frac{8\eta}{3}; \text{ when } \eta > \frac{\pi^2}{6}, I''_j \leqslant \frac{4\pi^2}{9}; \qquad (44.14)$$

the equality sign applying in the first case when $\eta \ll 1$ and in the second case when $\eta \gg 1$.

Consequently,

$$l_{ph.\overline{l}} = \frac{27 M a k_B \theta a \left(E_k\right)}{16\pi\beta'C^2\varepsilon^2\hbar^2 (ak)^4} \left(\frac{\theta}{T}\right)^2 \left(\frac{dE}{dk}\right)^2. \qquad (44.15)$$

As E_k increases from zero to infinity, the coefficient $\alpha(E_{k'})$ varies from 1 to 0.5.

It is interesting to compare these mean free paths with the mean free paths l_{ph} which are due to the normal phonon scattering. According to the formulas (15.29), (15.30), and (15.31), for metals at $T \gg \theta$

$$l_{ph} = \frac{M k^2 a^3 k_B \theta}{\pi^3 \hbar^2 C^2} \cdot \frac{\theta}{T} \left(\frac{dE}{dk}\right)^2, \qquad (44.16)$$

and at $T \ll \theta$

$$l_{ph} = \frac{M k^2 a^3 k_B \theta}{4\pi^3 124 . 4\hbar^2 C^2} \left(\frac{\theta}{T}\right)^5 \left(\frac{dE}{dk}\right)^2. \qquad (44.17)$$

For semiconductors at any value of T

$$l_{ph} = \frac{9 \left(\frac{4\pi}{3}\right)^{2/3} M a k_B \theta}{16\pi\hbar^2 C^2 (ak)^2} \cdot \frac{\theta}{T} \left(\frac{dE}{dk}\right)^2. \qquad (44.18)$$

Consequently, for metals at $T \gg \theta$

$$\frac{l_{ph.l}}{l_{ph}} = \frac{9\pi^4}{32\beta (ak)^6 \varepsilon^2} \simeq \frac{9\pi}{(ak)^6 \varepsilon^2}, \qquad (44.19)$$

and at $T \ll \theta$

$$\frac{l_{ph.l}}{l_{ph}} = \frac{9 \cdot 124.4\pi^2}{2\,(ak)^6\beta'\varepsilon^2} \left(\frac{T}{\theta}\right)^3 \simeq \frac{5600}{(ak)^6\,\varepsilon^2} \left(\frac{T}{\theta}\right)^3. \qquad (44.20)$$

For semiconductors at $T \gg \theta$

$$\frac{l_{ph.l}}{l_{ph}} = \frac{\pi^2}{\left(\frac{4\pi}{3}\right)^{2/3}\beta\,(ak)^2\,\varepsilon^2} \simeq \frac{2}{(ak)^2\,\varepsilon^2}, \qquad (44.21)$$

and at $T \ll \theta$

$$\frac{l_{ph.l}}{l_{ph}} = \frac{3\alpha}{\left(\frac{4\pi}{3}\right)^{2/3}\beta'\,(ak)^2\,\varepsilon^2} \cdot \frac{\theta}{T} \simeq \frac{1}{(ak)^2\,\varepsilon^2} \cdot \frac{\theta}{T}. \qquad (44.22)$$

In metals near the limit of population, ak is approximately equal to π. Therefore, according to Eqs. (44.19) and (44.20), $l_{ph.\,l}$ (both at high and very low temperatures) may be comparable with l_{ph} or even much smaller than it. In the latter case, the phonon−liquid scattering will be the dominant process. In semiconductors, $ak < 1$ for the majority of electrons, and the lower the temperature, the lower is the quantity. Therefore, according to Eqs. (44.21) and (44.22), $l_{ph.\,l} \gg l_{ph}$. At high temperatures, the phonon−liquid scattering is a small correction to the normal thermal scattering, and at low temperatures it can be neglected altogether.

§45. Change in the Electron Distribution Function Under the Action of the Optical-Mode Vibrations

We shall extend the calculations given in the present chapter to the case of ionic liquids exhibiting electronic conduction, in which, as in ionic crystals, the electrons are scattered mainly on the optical-mode vibrations and the related dipole polarization.

All the main ideas and representations are no different from those obtaining in the case of electron scattering on the acoustical vibrations, and therefore we shall present only very briefly the calculations which differ somewhat in the optical-mode vibration case.

Equation (41.3) remains valid, but in contrast to the acoustical case, we may approximately assume that all the optical oscillators vibrate at the same frequency ω_0.

According to Fröhlich [5], the perturbation energy in the interaction of electrons with the optical-mode lattice vibrations is:

$$U = -\frac{4\pi i e^2}{2a^3 q \sqrt{2G}} \sum_q (a_q e^{i q \cdot r} - a_q^* e^{-i q \cdot r}), \qquad (45.1)$$

where it is assumed that the electrons interact only with the longitudinal vibrations since the dipole polarization appears only for the longitudinal waves. We have established earlier that in liquids the electrons are scattered both by the longitudinal and transverse acoustical-mode vibrations. Similar scattering occurs also on the transverse optical-mode vibrations, and this scattering is of the same nature as on the acoustical waves; however, its contribution is relatively small and we shall neglect it, just as in ionic crystals it is usual to neglect the scattering of the electrons on the acoustical-mode vibrations.

Consequently, in Eqs. (41.3) and (45.1) we need include only the longitudinal phonons, and we need not sum over the phonon polarization directions, in contrast to the calculations in the preceding chapters.

After substituting Eq. (45.1) into Eq. (41.3), it is necessary to integrate with respect to the electron coordinates r and the oscillator coordinates a_q.

The integration with respect to a_q presents no special difficulties and yields the factors $\sqrt{\dfrac{\hbar}{2M\omega_0}} N_q$ or $\sqrt{\dfrac{\hbar}{2M\omega_0}} (N_q + 1)$, respectively, for the absorption and emission of a phonon with a wave number **q**.

The integral with respect to the electron coordinates can be reduced to the form:

$$K^{\pm} = \frac{1}{G} \sum_n e^{i(k \pm q - k') \cdot R_n} \int_1 e^{i(k \pm q - k') \cdot r'} U_k(r') U_{k'}(r') dr, \quad (45.2)$$

which is similar to the expression (42.5).

Therefore, we can repeat the remarks made in § 42. Two cases are possible.

1. If the law of conservation of momentum is obeyed by the electron—phonon system, i.e., $k \pm q = k'$, then the calculations to second-order small quantities do not differ from the calculation for crystals, and we may use the formulas which have been obtained for solids.

2. If $k \pm q \neq k'$, then for liquids (in contrast to crystals) $K^{\pm}$ is not equal to zero, but is calculated as a sum of random quantities. This gives rise to additional phonon—liquid scattering of the electrons.

We shall calculate the electron mean free path $l_{ph, l}$ which is governed by this scattering.

In calculating an integral over a unit cell, which occurs in Eq. (45.2), we shall take outside the integral symbol the average value of the product $U_k U_{k'}^*$, which is equal to $1/a^3$, assuming that over most of the unit cell this product is almost constant. This is true for metals (see p. 79 [2]), while for ionic substances this assumption may be regarded as the limiting case.

The integral remaining after these operations is calculated for one unit cell in the form of a cube with edge a, taking r' from the center of the cube. The ξ_α axes are directed along the cube edges and for brevity we shall use $\mathbf{k} \pm \mathbf{q} - \mathbf{k}' = \mathbf{q}'$, so that

$$\frac{1}{a^3} \int_1 e^{i\mathbf{q}' \cdot \mathbf{r}'} dr = \prod_{\alpha=1}^3 \frac{1}{a} \int_{-\frac{a}{2}}^{\frac{a}{2}} e^{iq'_\alpha \xi_\alpha} = \prod_{\alpha=1}^3 \frac{\sin \frac{q'_\alpha a}{2}}{\frac{q'_\alpha a}{2}} \simeq 1 - \frac{a^2 q'^2}{24} . \quad (45.3)$$

The above expression has been obtained by expanding in series the sines and dropping all powers of aq' higher than the second. In this operation, as well as in taking the average value of $U_k U_{k'}^*$ outside the integral symbol, we exaggerate the term in Eq. (45.3) which gives a correction to unity.

Substituting Eqs. (45.1) and (45.3), as well as the results of integration with respect to the oscillator coordinates, into Eq. (42.3) and integrating this equation with respect to time, we obtain expressions similar to Eqs. (42.18) and (42.19):

$$|c(\mathbf{k}', N_q + 1, t)|^2 = \frac{e^4 \varepsilon^2}{G^2 a^6 M \omega_0 \hbar} \Omega (E_{k'} - E_k + \hbar \omega_0) \times$$

$$\times \frac{N_q + 1}{q^2} \left(1 - \frac{a^2 q'^2}{12} \right); \quad (45.4)$$

$$|c(\mathbf{k}', N_q - 1, t)|^2 = \frac{e^4 \varepsilon^2}{G^2 a^6 M \omega_0 \hbar} \Omega (E_{k'} - E_k - \hbar \omega_0) \times$$

$$\times \frac{N_q}{q^2} \left(1 - \frac{a^2 q'^2}{12} \right); \quad (45.5)$$

here $\Omega(x)$ is given by Eq. (42.20).

We have dropped the term with the fourth power of q'. In order to obtain the change in the electron distribution function due to scattering, we shall multiply Eqs. (45.4) and (45.5) by the probability of reaching the initial state, and the probability of the final state being unoccupied, and we shall sum over all initial values of the electron wave number $\mathbf{k}$ and, independently, over all values of the phonon wave number $\mathbf{q}$. Consequently, we obtain

$$f(\mathbf{k}', t) - f(\mathbf{k}', 0) = \frac{e^4 \varepsilon^2}{G^2 a^6 M \omega_0 \hbar} \sum_{\mathbf{k}} \sum_{\mathbf{q}} \left(1 - \frac{a^2 q'^2}{12}\right) \times$$

$$\times \{ \mathfrak{Q}(E_{\mathbf{k}'} - E_{\mathbf{k}} + \hbar\omega_0)[f(\mathbf{k})(1 - f(\mathbf{k}'))(N_{\mathbf{q}} + 1) - f(\mathbf{k}')(1 - f(\mathbf{k}))N_{\mathbf{q}}] +$$

$$+ \mathfrak{Q}(E_{\mathbf{k}'} - E_{\mathbf{k}} - \hbar\omega_0) \times$$

$$\times [f(\mathbf{k})(1 - f(\mathbf{k}'))N_{\mathbf{q}} - f(\mathbf{k}')(1 - f(\mathbf{k})(N_{\mathbf{q}} + 1)] \}.$$

$$(45.6)$$

§ 46. Electron Mean Free Path Associated with the Phonon — Liquid Scattering in the Case of Optical-Mode Vibrations

In the $\mathbf{k}$ space, we shall introduce a spherical system of coordinates k, θ, Φ with the axis along k_x and in the $\mathbf{q}$ space use a different spherical system q, ϑ, φ with its axis along $\mathbf{k} - \mathbf{k}'$; we shall replace the summation over $\mathbf{k}$ and $\mathbf{q}$ by integration using the formulas:

$$\left.\begin{aligned}
\sum_{\mathbf{k}} F(\mathbf{k}) &= \mathfrak{Q}_0 \frac{G}{(2\pi)^3} \int k^2 dk \sin\theta d\theta d\Phi F(\mathbf{k}); \\
\sum_{\mathbf{q}} F(\mathbf{q}) &= \mathfrak{Q}_0 \frac{G}{(2\pi)^3} \int q^2 dq \sin\vartheta d\vartheta d\varphi F(\mathbf{q});
\end{aligned}\right\} \qquad (46.1)$$

here Ω_0 is the volume of one unit cell.

As before, we shall assume that the phonons are in thermal equilibrium and the electrons depart little from thermal equilibrium due to the application of an external electric field acting along the x axis. In this case,

$$N_{\mathbf{q}} = \frac{1}{e^{\frac{\hbar\omega_0}{k_B T}} - 1}; \quad f = f_0 + k_x f_1(E); \quad f_0 = \frac{1}{e^{\frac{E - \mu}{k_B T}} + 1}; \quad (46.2)$$

$f_1(E)$ is a small correction, the square of which we shall neglect; μ is the Fermi potential. The energy E is assumed to depend only on the modulus of the electron wave number (the isotropic approximation). $N_{\mathbf{q}}$ is found to be independent of $\mathbf{q}$, and, therefore, we shall omit the subscript $\mathbf{q}$.

Integration with respect to φ gives the factor 2π. By definition

$$q'^2 = (\mathbf{k} - \mathbf{k}' \pm \mathbf{q})^2 = (\mathbf{k} - \mathbf{k}')^2 + q^2 \pm 2q \,|\,\mathbf{k} - \mathbf{k}'\,| \cos\vartheta. \ (46.3)$$

On integration with respect to ϑ, the term with $\cos \vartheta$ disappears and the remaining terms are multiplied by 2. Integration with respect to q then gives

$$\int_0^{q_0} \left(1 - \frac{a^2 q'^2}{12}\right) dq = q_0 - \frac{a^2(k-k')^2 q_0}{12} - \frac{a^2 q_0^3}{36}. \qquad (46.4)$$

In semiconductors, $ak \ll 1$ for the great majority of electrons in the conduction band, and therefore we can neglect the second term in Eq. (46.4). In this case, using $q_0 = \frac{2\pi}{a} \sqrt[3]{\frac{3}{4\pi}}$, we obtain

$$\int_0^{q_0} \left(1 - \frac{a^2 q'^2}{12}\right) dq \simeq 0.57 q_0 = \frac{0.57 \cdot 2\pi}{a} \sqrt[3]{\frac{3}{4\pi}}, \qquad (46.5)$$

i.e., the inclusion of the factor $e^{i\mathbf{q} \cdot \mathbf{r}}$ in the integral over a unit cell gives a numerical coefficient of 0.57. If the calculation is made more exact, this coefficient will lie closer to 1.

Integration with respect to Φ yields the factor 2π; after substituting Eqs. (46.1), (46.2), (46.5), and the values of the integrals with respect to φ, ϑ and Φ, in Eq. (45.6) and differentiating with respect to time, we obtain

$$\left(\frac{\partial f}{\partial t}\right)_{\text{ph.}l} = \frac{0.57 \sqrt[3]{\frac{3}{4\pi}} e^4 \varepsilon^2}{4\pi^3 M \hbar \omega_0 a} \cdot \frac{\partial}{\partial t} \left\{ \int_{k'}^{k_{max}} k^2 dk \int_0^\pi \sin\theta d\theta \Omega (E_{k'} - E_k + \hbar\omega_0) \times \right.$$

$$\times [k_x f_1(E_k)(1 + N - f_0(E_{k'})) - k'_x f_1(E_k)(N + f_0(E_{k'}))] +$$

$$+ \int_{k_{min}}^{k'} k^2 dk \int_0^\pi \sin\theta d\theta \Omega (E_{k'} - E_k - \hbar\omega_0) \times$$

$$\left. \times [k_x f_1(E_k)(N + f_0(E_{k'})) - k'_x f_1(E_{k'})(1 + N - f_0(E_k))] \right\}. $$

$$(46.6)$$

On integration with respect to θ, the terms with $k_x = k \cos\theta$ drop out and the terms with $k'_x = k' \cos\theta'$ are multiplied by 2. Integrating with respect to k, we use the δ nature of the function Ω. All the factors, except Ωk, are placed outside the integral; assuming that $E_k = E_{k'} \pm \hbar\omega_0$, we extend the integration limit to infinity and use the fact that in a semiconductor

$$E_k = \frac{\hbar^2 k^2}{2m^*}, \quad E_{k'} = \frac{\hbar^2 k'^2}{2m^*}, \qquad (46.7)$$

where m* is the effective electron mass. Then integration with respect to k gives

$$\int_{-\infty}^{\infty} \Omega \left(E_{k'} - E_k \pm \hbar\omega_0 \right) k\,dk = \frac{2\pi m^*}{\hbar}\, t. \qquad (46.8)$$

Substituting the expression for $(\partial f/\partial t)_{ph.\,l}$ into the transport equation, assuming that an electric field E_x is applied along the x axis, and solving the resultant equation for $f_1(k')$, we obtain

$$f_1 = -\frac{\pi^2 M \hbar^2 \omega_0 a e E_x}{0.57 \sqrt[3]{\frac{3}{4\pi}}\, e^4 m^* \varepsilon^2 I} \cdot \frac{df_0}{dE} \cdot \frac{dE}{dk}, \qquad (46.9)$$

where

$$I = \sqrt{2m^*(E + \hbar\omega_0)}\,[N + f_0(E + \hbar\omega_0)] +$$

$$+ \sqrt{2m^*(E - \hbar\omega_0)}\ 1 + N - f_0(E - \hbar\omega_0)]. \qquad (46.10)$$

Hence, the electron mean free path associated with the phonon−liquid scattering on the optical-mode vibrations is

$$l_{ph.\bar{l}} = -\frac{k f_1}{e E_x \dfrac{df_0}{dE}} = \frac{\pi^2 M \hbar^2 \omega_0 a}{0.57 \sqrt[3]{\dfrac{3}{4\pi}}\, e^4 m^* \varepsilon^2 I} \cdot \frac{dE}{dk}. \qquad (46.11)$$

Using Eqs. (46.2), (46.7), (46.10), and (46.11), the mean free path may be calculated in closed form at any temperature but it is quite a complicated function of the temperature T and the electron energy E. The expression for I may be simplified considerably at high and low temperatures, separated by the Debye temperature

$$\theta = \frac{\hbar\omega_0}{k_B}. \qquad (46.12)$$

At $T \gg \theta$, we have $N \simeq k_B/\hbar\omega_0$, for the majority of electrons $E \gg \hbar\omega_0$, for semiconductors $f_0 \ll N$ and

$$I \simeq \frac{2k_B T}{\hbar\omega_0}\, \hbar k. \qquad (46.13)$$

Substituting Eq. (46.13) into Eq. (46.11), we obtain after making some simple transformations

$$l_{ph.l} = \frac{a\pi^2}{0.57 \sqrt[3]{\frac{3}{4\pi}} \, \epsilon^2} \cdot \frac{M}{m^*} \left(\frac{\hbar\omega_0}{\frac{e^2}{a}}\right)^2 \frac{E}{k_B T} \cdot \frac{1}{(ka)^2} \cdot \qquad (46.14)$$

At $T \ll \theta$,

$$N \simeq e^{\frac{-\hbar\omega_0}{k_B T}} \ll 1,$$

$$f_0(E + \hbar\omega_0) \simeq e^{\frac{\mu - E - \hbar\omega_0}{k_B T}} \ll 1, \quad \sqrt{2m^*(E + \hbar\omega_0)} \simeq \sqrt{2m^*\hbar\omega_0} \, .$$

The second term vanishes in Eq. (46.10) because the energy of the great majority of electrons is insufficient for phonon emission.

Consequently,

$$I \simeq \sqrt{2m^*\hbar\omega_0} \; e^{\frac{\hbar\omega_0}{k_B T}} \left(1 + e^{\frac{\mu - E}{k_B T}}\right) \simeq e^{-\frac{\hbar\omega_0}{k_B T}} \sqrt{2m^*\hbar\omega_0} \, , \quad (46.15)$$

and, after making some slight transformations, we obtain

$$l_{ph.l} = \frac{2a\pi^2}{0.57 \sqrt[3]{\frac{3}{4\pi}} \, \epsilon^2} \cdot \frac{M}{m^*} \cdot \frac{\hbar\omega_0 E}{\left(\frac{e^2}{a}\right)^2} \cdot \frac{1}{(ka)^2} e^{\frac{\hbar\omega_0}{k_B T}} \sqrt{\frac{E}{\hbar\omega_0}} \, . \quad (46.16)$$

We shall compare the expressions obtained for the mean free path $l_{ph.l}$ with those for the mean free path associated with the normal phonon scattering. According to Davydov and Shmushkevich [see Eqs. (3.97) and (3.98) in [6], in ionic crystals for the ion charge $z = 1$ at $T \gg \theta$ we have

$$l_{ph} = \frac{a}{2\pi} \cdot \frac{M}{m^*} \left(\frac{\hbar\omega_0}{\frac{e^2}{a}}\right)^2 \frac{E}{k_B T} \, , \qquad (46.17)$$

and at $T \ll \theta$

$$l_{ph} = \frac{a}{2\pi} \cdot \frac{M}{m^*} \left(\frac{\hbar\omega_0}{\frac{e^2}{a}}\right)^2 e^{\frac{\hbar\omega_0}{k_B T}} \sqrt{\frac{E}{\hbar\omega_0}} \, . \qquad (46.18)$$

Consequently, at $T \gg \theta$

$$\frac{l_{ph.l}}{l_{ph}} = \frac{2\pi^3}{0.57 \sqrt[3]{\frac{3}{4\pi}} (ka)^2 \varepsilon^2} \simeq \frac{180}{(ka)^2 \varepsilon^2} \gg 1, \qquad (46.19)$$

and at $T \ll \theta$

$$\frac{l_{ph.l}}{l_{ph}} = \frac{4\pi^3}{0.57 \sqrt[3]{\frac{3}{4\pi}} \varepsilon^2} \cdot \frac{E}{\hbar \omega_0} \cdot \frac{1}{(ka)^2} \simeq \frac{360}{(ka)^2} \cdot \frac{E}{\hbar \omega_0 \varepsilon^2} \gg 1. \quad (46.20)$$

Thus in ionic semiconductors, as in covalent semiconductors, when the scattering of the electrons on the optical-mode vibrations is included, the phonon—liquid scattering is small compared with the usual thermal scattering, which is calculated for liquid semiconductors in exactly the same way as for crystals.

Chapter VIII

ELECTRICAL CONDUCTIVITY,
THERMAL CONDUCTIVITY,
THERMOELECTRIC POWER, HALL COEFFICIENT,
AND NERNST COEFFICIENT
OF AMORPHOUS SUBSTANCES
EXHIBITING ELECTRONIC CONDUCTION [1]

§47. General Formulas

In preceding chapters, we investigated the general features of the scattering of electrons in amorphous conductors and calculated the electron mean free path allowing for the scattering due to the departures from long-range order and for the phonon—liquid scattering. However, experimentally we determine not the mean free path, but the electrical conductivity, Hall coefficient, and similar quantities. Since the dependence of the mean free path on the electron energy is different for amorphous conductors and crystals, we may expect a different temperature dependence of the electrical conductivity and other properties. The purpose of the present chapter is to determine this temperature dependence.

We shall take into account all types of electron scattering in a liquid, introducing a general electron mean free path l. To determine l, it is necessary to add the reciprocals of the mean free paths associated with different scattering mechanisms:

$$\frac{1}{l} = \frac{1}{l_l} + \frac{1}{l_{ph}} + \frac{1}{l_{ph.l}} + \frac{1}{l_D}. \tag{47.1}$$

When electrons are scattered on thermal vibrations in metals at high temperatures $T \gg \theta$ (where θ is the Debye temperature) and on thermal vibrations in semiconductors at any temperature, there is a universal electron mean free

179

path which appears in all the transport processes. The universal mean free path l always exists for scattering in liquids. Thus, with the exception of metals at $T < \theta$ (this case is of no interest to us because at low temperatures metals are in the crystalline state), we can regard the electron mean free path as a universal quantity.

To simplify our formulas, we shall use the effective mass approximation, which is satisfactory for semiconductors but not as good for metals. However, we are interested only in the temperature dependence of the various parameters and not in their absolute values, therefore we can satisfactorily use the effective mass approximation even for metals. With these assumptions we can calculate the transport coefficients using the free-electron theory.

For fixed values of l the formulas of this theory have been given very fully by Brillouin (Chap. VII in [2]). Brillouin uses the notation

$$u = \frac{E}{k_B T} = \frac{m^* v^2}{2k_B T},$$ (47.2)

where E is the electron energy; k_B is Boltzmann's constant; m^* is the effective electron mass; and v is the electron velocity. Moreover,

$$V_{k-1} = \frac{1}{k!} \int_0^\infty f_0 \frac{\partial (lu^k)}{\partial u} \, du = -\frac{1}{k!} \int_0^\infty \frac{\partial f_0}{\partial u} lu^k du;$$ (47.3)

$$K_{jk} = \int_0^\infty f_0 \frac{\partial}{\partial v} \cdot \frac{l^j v^k}{1 + \frac{\mu_H^2 l^2}{v^2}} \, dv; \quad K'_{jk} = \int_0^\infty \frac{\partial f_0}{\partial T} \cdot \frac{l^j v^k}{1 + \frac{\mu_H^2 l^2}{v^2}} \, dv;$$ (47.4)

here

$$\mu_H = \frac{eH}{m^* c},$$ (47.5)

f_0 is the equilibrium Fermi distribution function; e is the electronic charge; H is the magnetic field; and c is the velocity of light. The transport coefficients are expressed in terms of the integrals given above. The electrical conductivity is

$$\sigma = \frac{16\pi e^2 m^*}{3h^3} k_B TV_0,$$ (47.6)

where h is Planck's constant.

The thermal conductivity is

$$\varkappa = \frac{32\pi m^*}{3} \left(\frac{k_B}{h} \right)^2 T^2 \left(3V_2 - \frac{2V_1^2}{V_0} \right).$$ (47.7)

The Thomson coefficient is

$$\sigma_T = \frac{k_B T}{e} \cdot \frac{d}{dT} \left(2 \frac{V_1}{V_0} - \tilde{\mu} \right),$$ (47.8)

where $\tilde{\mu} = e\mu / k_B T$, and μ is the Fermi potential.

The Peltier coefficient is given by

$$\Pi_{12} = \frac{k_B T}{e} \left| 2 \frac{V_1}{V_0} - \tilde{\mu} \right|_1^2.$$ (47.9)

The symbol $\left| \ \right|_1^2$ means that we must take the difference of the expressions between the two verticals for conductors 2 and 1.

The thermoelectric power is

$$\alpha_T = \frac{k_B}{e} \left| -2 \frac{V_1}{V_0} + \tilde{\mu} \right|_1^2 = - \frac{\Pi_{12}}{T}.$$ (47.10)

The isothermal electrical conductivity in a magnetic field is given by

$$\mathfrak{I}_{Hi} = \frac{8\pi e}{3} \left(\frac{m^*}{h} \right)^3 \frac{e}{m^*} \left[K_{12} + \mu_H^2 \frac{K_{21}^2}{K_{12}} \right] =$$
$$= \sigma_0 \frac{K_{12}(H)}{K_{12}(0)} \left(1 + \mu_H^2 \frac{K_{21}^2}{K_{12}^2} \right).$$ (47.11)

The isothermal Hall coefficient is

$$R_i = \frac{e}{cm^* \sigma_{Hi}} \cdot \frac{K_{21}}{K_{12}} = \frac{3}{8\pi enc} \left(\frac{h}{m^*} \right)^3 \frac{K_{21}}{K_{12}^2 + \mu_H^2 K_{21}^2}.$$ (47.12)

The isothermal Nernst coefficient is

$$Q_i = \frac{E_y}{H_z \frac{\partial T}{\partial x}} - \frac{K_{12} K_{22}' - K_{21} K_{13}'}{K_{12}^2 + \mu_H^2 K_{21}^2};$$ (47.13)

E_y is the y component of the electric field appearing in a magnetic field H_z when a temperature gradient exists along the x axis. Similar expressions are obtained for the adiabatic Hall and Nernst coefficients and for the Ettings-hausen and Righi—Leduc coefficients, but are not given here because they are very cumbersome.

The next step consists of the substitution of the expression for the mean free path l into the formulas for actual cases, and the calculation of the integrals (47.3) and (47.4) and then of the coefficients given by Eqs. (47.6)-(47.13). We shall consider separately metals at high temperatures, covalent semiconductors, and ionic semiconductors with weak coupling between the lattice vibrations and the electron motion. The strong coupling case (polarons) in ionic amorphous semiconductors has not yet been considered by anybody.

§ 48. Liquid Metals At $T \gg \theta$

The integrals (47.3) and (47.4) for metals are calculated by expanding them in series of powers of $1/\tilde{\mu}$, where $\tilde{\mu}$ is taken from the lower edge of a semifilled band; in metals $\tilde{\mu} \gg 1$. Using Brillouin's formulas [2]:

$$V_{k-1} = \frac{1}{k!} \left\{ lu^k + \frac{\pi^2}{6} \times \right.$$

$$\times \left[k(k-1)u^{k-2}l + 2ku^{k-1}\frac{dl}{du} + u^k \frac{d^2l}{du^2} \right] + \dots \left. \right\}; \quad (48.1)$$

$$K_{jk} = \left(\frac{2k_B T}{m^*} \right)^{\frac{k}{2}} \left[\frac{l^j u^{\frac{k}{2}}}{1 + \frac{\gamma}{u}} + \frac{\pi^2}{6} \cdot \frac{\partial^2}{\partial u^2} \left(\frac{l^j u^{\frac{k}{2}}}{1 + \frac{\gamma}{u}} \right) + \dots \right]; \quad (48.2)$$

$$K'_{jk} = \frac{1}{2T} K_{j,k+1} - \frac{k_B}{m^*} \left(\tilde{\mu} + \frac{\pi^2}{6\tilde{\mu}} \right) K_{j,k-1}; \quad (48.3)$$

where

$$\gamma = \frac{\mu_H^2 l^2 m^*}{2k_B T} . \quad (48.4)$$

After the substitution of these expressions into Eqs. (47.6)-(47.13), we shall retain only the largest terms which give a nonzero result. Thus, in σ, σ_{Hi}, and R_i, we need retain only the first term in the expressions (48.1)-(48.3). For all the other properties related to thermal motion, this zeroth-order approximation gives a zero result and, therefore, we must retain the second term.

To determine the mean free path l at $T > \theta$, we shall substitute the expressions (38.17), (39.8), (44.5), and (44.16) into Eq. (47.1), to obtain

$$\frac{1}{l} = \frac{k^2}{\left(\frac{dE}{dk}\right)^2}\left(A_1 + B_1 k^2 + \frac{D_1}{k^4}\right); \tag{48.5}$$

here

$$A_1 = \varepsilon^2 A_1' + A_1''; \quad A_1 = \frac{c\hbar^4}{8\pi a m^2} + \frac{\overline{|V_n'|^2} a^3}{\pi}; \left.\begin{array}{c} \\ \\ \end{array}\right\}$$

$$A_1'' = \frac{\overline{V_D^2} a^3}{\pi}; \tag{48.6}$$

$$B_1 = \left(B_1' + B_1'' \frac{T}{\theta}\right)\varepsilon^2 - B_1'''; \quad B_1' = \frac{c_1 \hbar^4 a}{8\pi m^2} - \frac{2\overline{|V_n'|^2} a^5}{9\pi};$$

$$B_1'' = \frac{32\beta C^2 \hbar^2 a^3}{9\pi M k_B \theta}; \quad B_1''' = \frac{2\overline{|V_D|^2} a^5}{9\pi}; \left.\right\} \tag{48.7}$$

$$D_1 = D_1' \frac{T}{\theta}; \quad D_1' = \frac{\pi^3 C^2 \hbar^2}{M a^3 k_B \theta}; \tag{48.8}$$

θ is the Debye temperature.

In the effective mass approximation, Eq. (48.5) becomes:

$$l = \frac{\hbar^4}{m^{*2}\left(A_1 + B_1 k^2 + \frac{D_1}{k^4}\right)}; \tag{48.9}$$

here A_1 and B_1 are coefficients which depend on the departure from order in a liquid metal, and D_1 at $T \gg \theta$ is a coefficient proportional to temperature; k is the electron wave number and m* is the effective electron mass.

Using the formulas (47.6)-(47.13) with Eqs. (48.1)-(48.3) and (48.9) substituted in them, we obtain expressions for σ, $\varkappa$, σ_T, α_T, and Π_{12}. The electrical conductivity is

$$\sigma = \frac{e^2 \hbar^3 k_F^2}{3\pi^2 m^{*2}\left(A_1 + B_1 k_F^2 + \frac{D_1}{k_F^4}\right)}; \tag{48.10}$$

k_F is the value of the electron wave number at the Fermi surface when $E = e\mu$. Since the electron density in a metal is

$$n = \frac{8\pi}{3} \cdot \frac{1}{(2\pi)^3} \, k_F^3 = \frac{k_F^3}{3\pi^2}, \qquad (48.11)$$

it follows that the electron mobility is

$$u_1 = \frac{e \, \hbar^3}{m^{*2} k_F \left(A_1 + B_1 k_F^2 + \dfrac{D_1}{k_F^4} \right)}. \qquad (48.12)$$

The other properties are

$$\varkappa = \frac{k_B^2 T \hbar^3 k_F^2}{9 m^* \left(A_1 + B_1 k_F^2 + \dfrac{D_1}{k_F^4} \right)}; \qquad (48.13)$$

$$\sigma_T = \frac{2\pi^2 k_B^2 m^* T}{3 e \hbar^3 k_F^2} \left[1 + \frac{d}{dT} \cdot \frac{T(2 D_1 - B_1 k_F^6)}{A_1 k_F^4 + B_1 k_F^6 + D_1} \right]; \qquad (48.14)$$

$$\alpha_T = - \frac{2 k_B^2 T}{3 e \hbar^2} \left| \frac{m^*}{k_F^2} \left(1 + \frac{2 D_1 - B_1 k_F^6}{A_1 k_F^4 + B_1 k_F^6 + D_1} \right) \right|_1^2; \qquad (48.15)$$

$$\Pi_{12} = -\alpha_T T. \qquad (48.16)$$

The expressions for R_i, σ_{Hi}, and Q_i in their general form are very cumbersome and, therefore, we shall restrict ourselves to weak magnetic fields for which $\gamma/\tilde{\mu} \ll 1$ so that in the expressions for R_i and Q_i we can neglect altogether the terms with $\gamma/\tilde{\mu}$, and in the expression for σ_{Hi} we need include only the term with the first power of $\gamma/\tilde{\mu}$. Then, the Hall coefficient is found to be independent of the mean free path; consequently, for liquid metals we have exactly the same expression as for solid metals. For σ_{Hi} and Q_i, we obtain the following expressions

$$\sigma_{Hi} = \sigma_0 \left[1 - \frac{\pi^2 k_B^2 T^2 \hbar^2 k_F^2 e^2 H^2}{3 m^{*2} c^2 \left(A_1 k_F^4 + B_1 k_F^6 + D_1 \right)^2} \right]; \qquad (48.17)$$

$$Q_i = \frac{\pi^2 k_B^2 T \hbar k_F \left(A_1 k_F^4 + 2 B_1 k_F^6 - D_1 \right)}{3 m^* \left(A_1 + B_1 k_F^2 + \dfrac{D_1}{k_F^4} \right)^2}. \qquad (48.18)$$

The expressions for R_i and σ_{Hi} which are valid in arbitrary magnetic fields can be obtained by substituting into the appropriate formulas (see pp. 237-288 in [2]) the expression (48.9) for the electron mean free path in a liquid metal. In order to obtain the explicit temperature dependences of the galvanomagnetic coefficients, we shall substitute Eqs. (48.6)-(48.8) into Eqs. (48.12)-(48.15) and (48.17)-(48.18) to obtain

$$u_1 = \frac{e\hbar^3}{m^{*2}k_F\left[f_1 + f_2T + \varepsilon^2(T)(f_3 + f_4T)\right]}; \qquad (48.19)$$

$$\varkappa = \frac{k_B\hbar^3k_F^2T}{9m^*\left[f_1 + f_2T + \varepsilon^2(T)(f_3 + f_4T)\right]}; \qquad (48.20)$$

$$\sigma_T = \frac{2\pi^2k_B^2m^*T}{3e\hbar^3k_F^2}\left\{1 + \frac{d}{dT}\cdot\frac{T\left[2f_2T - \varepsilon^2(T)(f_5 + f_4T) - f_6\right]}{f_1 + f_2T + \varepsilon^2(T)(f_3 + f_4T)}\right\}; \quad (48.21)$$

$$\alpha_T = -\frac{2k_B^2T}{3e\hbar^2}\left|\frac{m^*}{k_F^2}\left[1 + \frac{2f_2T - \varepsilon^2(T)(f_5 + f_4T) - f_6}{f_1 + f_2T + \varepsilon^2(T)(f_3 + f_4T)}\right]\right|_1^2; \quad (48.22)$$

$$\sigma_{Hi} = \sigma_0\left\{1 - \frac{\pi^2k_B^2\hbar^2e^2H^2T^2}{3m^{*2}c^2k_F^6\left[f_1 + f_2T + \varepsilon^2(T)(f_3 + f_4T)\right]^2}\right\}; \quad (48.23)$$

$$Q_i = \frac{\pi^2k_B^2\hbar k_F^5\left[f_1 + f_6 - f_2T + \varepsilon^2(T)(f_3 + f_5 + 2f_4T)\right]}{3m^*\left[f_1 + f_2T + \varepsilon^2(T)(f_3 + f_4T)\right]^2}. \quad (48.24)$$

Here $f_1, f_2, \ldots, f_6$ are temperature-independent constants which must be found by experiment. They are related to the constants $A'_1, \ldots, D'_1$ by:

$$\left.\begin{array}{l} f_1 = A''_1 + B''_1; \quad f_2 = \dfrac{D'_1}{k_F^4\theta}; \quad f_3 = A_1 + B'_1k_F^2; \\[3mm] f_4 = \dfrac{B''_1k_F^2}{\theta}; \quad f_5 = B'_1k_F^2; \quad f_6 = \dfrac{2B'''_1k_F^2}{\theta}. \end{array}\right\} \qquad (48.25)$$

The constants f_1 and f_6 represent the scattering on defects, f_2 represents the scattering on phonons, f_3 and f_5 represent liquid scattering (including the term V'), and f_4 represents phonon−liquid scattering.

The dependence $\varepsilon(T)$ should be found experimentally; in accordance with Eq. (34.11) it is given by the curve shown in Fig. 16 which represents the temperature dependence of the structural diffusion coefficient.

For this dependence, Glauberman [3] suggested a semi-empirical formula:

$$\varepsilon^2(T) = \varepsilon_m^2 e^{-\frac{Q}{R(T - T_{mp})}}, \qquad (48.26)$$

where ε_m is the maximum value of ε; Q is the energy needed to break up the lattice; R is the gas constant; T_{mp} is the melting point. Thus, according to Glauberman [3], for liquid tin $\varepsilon_m^2 = 0.0065$, Q = 36 kcal/mole, and for bismuth $\varepsilon_m^2 = 0.0043$, Q = 12.6 kcal/mole.

According to Fig. 16 and Eq. (48.26), ε^2 increases at temperatures several hundred degrees above the melting point and then remains constant. However, on approach to the critical temperature, the degree of disorder should again increase sharply and Eq. (48.26) should cease to be valid.

It follows that the temperature dependence of the thermogalvanomagnetic coefficients of liquid metals is complex; unfortunately, there are too few accurate experimental data to make a comparison with the theoretical temperature dependences.

§49.   Amorphous Covalent Semiconductors

In the case of semiconductors, we may assume that

$$f_0 = e^{\tilde{\mu} - u}, \tag{49.1}$$

where $\tilde{\mu}$ is the dimensionless Fermi potential.

Substituting Eqs. (38.17), (39.8), and (44.18) into Eq. (47.1) and neglecting the phonon—liquid scattering, we obtain, by virtue of the final conclusion of § 44

$$l = \frac{1}{A + Bu}; \tag{49.2}$$

here

$$A = \varepsilon^2 A' + A'' \frac{T}{\theta} + A'''; \quad B = \varepsilon^2 B' \frac{T}{\theta} - B'' \frac{T}{\theta}; \tag{49.3}$$

$$A' = \frac{c}{8\pi a} \left(\frac{m^*}{m}\right)^2 + \frac{m^{*2} \overline{|V_n'|^2} a^3}{\pi \hbar^4};$$

$$A'' = \frac{16\pi C^2 m^{*2}}{9 \left(\frac{4\pi}{3}\right)^{2/3} M a \hbar^2 k_B \theta}; \quad A''' = \frac{m^{*2} \overline{|V_D|^2} a^3}{\pi \hbar^4}; \tag{49.4}$$

$$B' = \frac{k_B \theta c_1 m^*}{4\pi\hbar^2} \left(\frac{m^*}{m}\right)^2 - \frac{2m^{*3} \overline{|V'_n|^2} a^5 k_B \theta}{9\pi\hbar^6} \; ;$$

$$B'' = \frac{2m^{*3} \overline{|V_D|^2} a^5 k_B \theta}{4\pi\hbar^6} .$$

$$(49.5)$$

Substituting Eqs. (49.1) and (49.2) into Eqs. (47.3) and (47.4), we can reduce the latter to exponential integrals E_i, but the expressions obtained are cumbersome and unsuitable for practical use. Therefore, we shall use the fact that for the great majority of electrons in the conduction band of a semiconductor $Bu \ll A$, which is easily proved using the expressions (49.3)-(49.5) for A and B.

The denominator in Eq. (49.2) is expanded in a series of powers of Bu/A, retaining only the zeroth and first-order terms. Similarly, we can calculate the integrals K_{jk}, considering only weak magnetic fields and including only terms of the first degree in $\mu_H^2 l^2/v^2$.

After integration we obtain

$$V_{k-1} = \frac{e^{\tilde{\mu}}}{A} [1 - \alpha (k+1)].$$

$$(49.6)$$

For $k = 2\nu + 1$

$$K_{jk} = \frac{e^{\tilde{\mu}}}{A^j} \left(\frac{2k_B T}{m^*}\right)^{\nu + \frac{1}{2}} \sqrt{\pi} \left[\frac{1 \cdot 3 \cdots (2\nu + 1)}{2^{\nu+1}} \right.$$
$$\left. - j\alpha \frac{1 \cdot 3 \cdots (2\nu + 3)}{2^{\nu+2}} - \beta \frac{1 \cdot 3 \cdots (2\nu - 1)}{2^\nu} \; ;\right.$$

$$(49.7)$$

for $k = 2\nu$

$$K_{jk} = \frac{e^{\tilde{\mu}}}{A^j} \left(\frac{2k_B T}{m^*}\right)^\nu [\nu! - j\alpha (\nu+1)! - \beta (\nu - 1)!];$$

$$(49.8)$$

here

$$\alpha = \frac{B}{A}, \quad \beta = \frac{m^* \mu_H^2}{2k_B T A^2} .$$

$$(49.9)$$

It is also easy to show that for semiconductors,

$$K'_{jk} = \frac{1}{2T} K_{j,\,k+1} + \frac{k_B T}{m^*} \cdot \frac{\partial \tilde{\mu}}{\partial T} K_{j,\,k-1}. \tag{49.10}$$

Substituting Eqs. (49.6)-(49.10), and (49.3) into Eqs. (47.6)-(47.13), we obtain

$$\sigma = \frac{16\pi e^2 m^*}{3h^3} k_B T \frac{A'\varepsilon^2 + (A'' - 2B'\varepsilon^2 - 2B'') \dfrac{T}{\theta}}{\left(A'\varepsilon^2 + A'' \dfrac{T}{\theta} + A''' \right)^2} e^{\tilde{\mu}}, \tag{49.11}$$

and since for semiconductors

$$n = \frac{2}{h^3} (2\pi m^* k_B T)^{3/2} e^{\tilde{\mu}}, \tag{49.12}$$

we have therefore

$$u_1 = \frac{2\sqrt{2}\, e}{3\sqrt{\pi k_B T m^*}} \cdot \frac{A'\varepsilon^2 + (A'' - 2B'\varepsilon^2 - 2B'') \dfrac{T}{\theta}}{\left(A'\varepsilon^2 + A'' \dfrac{T}{\theta} + A''' \right)^2}, \tag{49.13}$$

$$\varkappa = \frac{4\sqrt{2}\, k_B^{3/2} \sqrt{T}}{3\sqrt{\pi}} \cdot \frac{A'\varepsilon^2 + (A'' - 4B'\varepsilon^2 - 4B'') \dfrac{T}{\theta}}{\left(A'\varepsilon^2 + A'' \dfrac{T}{\theta} + A''' \right)^2} n, \tag{49.14}$$

$$\sigma_T = \frac{k_B T}{e} \cdot \frac{d}{dt} \left[2 \frac{A'\varepsilon^2 + (A'' - B'\varepsilon^2 - 2B'') \dfrac{T}{\theta}}{A'\varepsilon^2 + A'' \dfrac{T}{\theta} + A'''} - \tilde{\mu} \right], \tag{49.15}$$

$$\alpha_T = \frac{k_B}{e} \left| \tilde{\mu} - 2 \frac{A'\varepsilon^2 + (A'' - B'\varepsilon^2 - B'') \dfrac{T}{\theta}}{A'\varepsilon^2 + A'' \dfrac{T}{\theta} + A'''} \right|_1^2, \tag{49.16}$$

$$\sigma_{Hi} = \sigma_0 \left[1 - \left(1 - \frac{\pi}{4} \right) \frac{e^2 H^2}{2k_B T m^* c^2 \left(A'\varepsilon^2 + A'' \dfrac{T}{\theta} + A''' \right)^2} \right], \tag{49.17}$$

$$R_i = \frac{3\pi}{8enc} \left[1 + \frac{(B'\varepsilon^2 + B'') \dfrac{T}{\theta}}{A'\varepsilon^2 + A'' \dfrac{T}{\theta} + A'''} - \frac{e^2 H^2}{2k_B T m^* c^2 \left(A'\varepsilon^2 + A'' \dfrac{T}{\theta} + A''' \right)^2} \right], \tag{49.18}$$

$$Q_i = \frac{1}{4} \sqrt{\frac{\pi k_B}{2m^* T}} \cdot \frac{1}{A'\varepsilon^2 + A''\frac{T}{\theta} A'''} \left[1 + \frac{(B'\varepsilon^2 + B'')\frac{T}{\theta}}{A'\varepsilon^2 + A''\frac{T}{\theta} + A'''} \right.$$

$$\left. - \left(3 + \frac{\pi}{4}\right) \frac{e^2 H^2}{2k_B T m^* c^2 \left(A'\varepsilon^2 + A''\frac{T}{\theta} + A'''\right)} \right]. \qquad (49.19)$$

For amorphous and glassy semiconductors, in which the degree of disorder is "frozen", ε = const, but for liquid semiconductors ε depends on temperature. We may assume that the nature of this dependence is the same as for metals, i.e., the expression (48.26) is valid over a range of temperatures above the melting point. The temperature dependences of the thermogalvanomagnetic coefficients for liquid semiconductors are complex. If we assume that A' = B' = 0 (these quantities represent the "liquid" scattering of electrons), we obtain formulas for crystalline semiconductors. Conversely, at sufficiently low temperatures, we may neglect the terms $A''T/\theta$, which describe the electron scattering on thermal vibrations and are small compared with A', and then our formulas give the "liquid" scattering. The terms A''' and B''' represent the scattering on defects.

§50. Amorphous Ionic Semiconductors

The electron mean free path of ionic semiconductors is different at high and low temperatures: substituting Eqs. (38.17), (39.8), (46.17), and (46.18) into Eq. (47.1) and again neglecting the phonon−liquid scattering, we have at $T \gg \theta$

$$l = \frac{1}{A + Bu + \dfrac{D}{u}}, \qquad (50.1)$$

and at $T \ll \theta$

$$l = \frac{1}{A + Bu + \dfrac{F}{\sqrt{u}}}. \qquad (50.2)$$

Here the coefficients A, B, A', A''', B', and B" have the same meanings as for covalent semiconductors, i.e., they are given by the expressions (49.3), (49.4), and (49.5):

$$D = \frac{2\pi m^* e^4}{Ma^3\hbar^3\omega_0} \; ; \quad F = F'e^{-\frac{\theta}{T}} \sqrt{\frac{\theta}{T}} \; ; \quad F' = \frac{2\sqrt{2}\,\pi m^{*3/2}e^4}{Ma^3\hbar^{7/2}\omega_0^2\sqrt{k_B\theta}} \quad .(50.3)$$

At $T \gg \theta$

$$A'' = \frac{2 \cdot 0.57\,\sqrt[3]{\dfrac{3}{4\pi}}\; am^{*2}\theta}{\pi^2 M\omega_0^2\hbar^4} \; , \tag{50.4}$$

but at $T \ll \theta$, $A^* = 0$; ω_0 is the frequency of the optical-mode vibrations and θ is the corresponding Debye temperature. The terms D/u and $F/\sqrt{u}$ represent the scattering of electrons on the optical-mode thermal vibrations.

The expressions for l are very complicated and therefore we have to use much rougher approximations than in the case of covalent semiconductors: we shall neglect completely the terms B'u, which were included in the linear approximation for covalent semiconductors. At high temperatures, the remaining terms A and D/u are, in general, comparable and they must be calculated exactly. At low temperatures, the phonon scattering in amorphous and glassy substances is small compared with the liquid scattering and it is sufficient to include the term F/√u only in the linear approximation. Therefore, we shall carry out separate calculations for high and low temperature conditions.

High-Temperature Region $T \gg \theta$

Substituting Eqs. (49.1) and (50.1) into Eq. (47.3) and assuming that $D/A = \alpha_1$, we obtain

$$V_{k-1} = \frac{e^{\tilde{\mu}}}{A}\left[1 - \alpha_1 + \ldots + (-)^k\frac{\alpha_1^k}{k!} + (-)^k\frac{\alpha_1^{k+1}}{k!}e^{\alpha_1}\,\mathrm{Ei}\,(-\alpha_1)\right],$$

$$\tag{50.5}$$

where Ei is an exponential integral.

Similarly, the substitution of Eqs. (49.1) and (50.1) into Eq. (47.4) for weak magnetic fields gives

$$K_{jk} = \frac{e^{\tilde{\mu}}}{A^j}\left(\frac{2k_BT}{m^*}\right)^{\frac{k}{2}}[I_{jk}(\alpha_1) - \beta I_{j+2,\,k-2}(\alpha_1)], \tag{50.6}$$

where

$$I_{jk}(\alpha_1) = \int_0^\infty e^{-u} \frac{u^{j+\frac{k}{2}}}{(u+a_1)^j} du \qquad (50.7)$$

is an integral which has to be tabulated by numerical integration, and β is given by Eq. (49.9). Substituting Eqs. (49.3), (50.3), and (50.4) into Eqs. (47.6)-(47.13), we obtain:

$$\sigma = \frac{16\pi e^2 m^*}{3h^3} k_B T \frac{e^{\tilde\mu}}{A} \left[1 - \alpha_1 - \alpha_1^2 e^{\alpha_1} \operatorname{Ei}(-\alpha_1)\right]; \qquad (50.8)$$

$$u_1 = \frac{2\sqrt{2}\,e}{3\sqrt{\pi k_B Tm^* A}} \left[1 - \alpha_1 - \alpha_1^2 e^{\alpha_1} \operatorname{Ei}(-\alpha_1)\right]; \qquad (50.9)$$

$$\varkappa = \frac{4\sqrt{2}\,k_B^{3/2}\sqrt{T}\,n}{3\sqrt{\pi}\,A} \left[\frac{1 - 2\alpha_1 + \alpha_1^2 - (3\alpha_1^2 + \alpha_1^3)\,e^{\alpha_1}\operatorname{Ei}(-\alpha_1)}{1 - \alpha_1 - \alpha_1^2 e^{\alpha_1}\operatorname{Ei}(-\alpha_1)}\right]; \qquad (50.10)$$

$$\sigma_T = \frac{k_B T}{e} \cdot \frac{d}{dt}\left[\frac{2 - \alpha_1 + \alpha_1^2 + \alpha_1^3 e^{\alpha_1}\operatorname{Ei}(-\alpha_1)}{1 - \alpha_1 - \alpha_1^2 e^{\alpha_1}\operatorname{Ei}(-\alpha_1)} - \tilde\mu\right]; \qquad (50.11)$$

$$\alpha_T = \frac{k_B}{e}\left|\tilde\mu - \left[\frac{2 - \alpha_1 + \alpha_1^2 + \alpha_1^3 e^{\alpha_1}\operatorname{Ei}(-\alpha_1)}{1 - \alpha_1 - \alpha_1^2 e^{\alpha_1}\operatorname{Ei}(-\alpha_1)}\right]\right|_1^2; \qquad (50.12)$$

$$\sigma_{Hi} = \sigma_0\left\{1 - \frac{e^2 H^2}{2k_B T A^2 m^* c^2}\left[\frac{I_{30}(\alpha_1) I_{12}(\alpha_1) - I_{21}^2(\alpha_1)}{I_{12}^2(\alpha_1)}\right]\right\}; \qquad (50.13)$$

$$R_i = \frac{3\sqrt{\pi}}{4enc\left[1 - \alpha_1 - \alpha_1^2 e^{\alpha_1}\operatorname{Ei}(-\alpha_1)\right]} \times$$

$$\times\left\{1 - \frac{e^2 H^2}{2k_B T A^2 m^* c^2}\left[\frac{I_{4,-1}}{I_{21}} - \frac{2I_{30}I_{12} - I_{21}^2}{I_{12}^2}\right]\right\}; \qquad (50.14)$$

$$Q_i = \frac{1}{A}\sqrt{\frac{k_B}{2m^* T}}\left\{I_{21}I_{14} - I_{12}I_{23} + \frac{e^2 H^2}{2k_B T A^2 m^* c^2} \times\right.$$

$$\left.\left[(I_{12}I_{28} - I_{21}I_{14})\frac{I_{21}^2}{I_{12}^2} + I_{30}I_{23} + I_{12}I_{41} - I_{21}I_{32} - I_{4,-1}I_{14}\right]\right\}. \qquad (50.15)$$

For brevity, we have omitted the argument of α_1 in I_{jk} in the last two expressions. In the derivation of Eq. (50.15), we have also used Eq. (49.10).

The values of ε, A, α_1, I_{jk} for amorphous and glassy semiconductors are independent of temperature and, therefore, the complex expressions in the square brackets of Eqs. (50.8)-(50.15) can be replaced by constants found directly from experiment. It is not, in fact, necessary to calculate the integrals (50.7). The temperature dependence of the transport coefficients is simple, and, in this approximation, it is exactly the same as for crystalline ionic semiconductors. However, in the case of true liquids the temperature dependence of these coefficients is very complex and, therefore, in contrast to the treatment in § 48 and 49, we shall not give the expressions for this dependence in explicit form.

Low-Temperature Region $T \ll \theta$

Substituting Eqs. (49.1) and (50.2) into Eqs. (47.3) and (47.4), assuming that $F/A = \alpha_2$ and expanding in series of powers of α_2 and β using only the linear terms (β has its previous meaning), we obtain:

$$V_{k-1} = \frac{e^{\tilde{\mu}}}{A}\left[1 - \alpha_2 \frac{1 \cdot 3 \cdots (2k-1)}{2^{k-1}} \sqrt{\pi}\right] ; \qquad (50.16)$$

for $k = 2\nu + 1$

$$K_{jk} = \frac{e^{\tilde{\mu}}}{A^j}\left(\frac{2k_B T}{m^*}\right)^{\nu + \frac{1}{2}} \sqrt{\pi}\left[\frac{1 \cdot 3 \cdots (2\nu + 1)}{2^{\nu+1}} - \right.$$
$$\left. - \frac{j\alpha_2 \nu !}{\sqrt{\pi}} - \beta \frac{1 \cdot 3 \cdots (2\nu - 1)}{2^{\nu}}\right] ; \qquad (50.17)$$

for $k = 2\nu$

$$K_{jk} = \frac{e^{\tilde{\mu}}}{A^j}\left(\frac{2k_B T}{m^*}\right)^{\nu}\left[\nu ! - j\alpha_2 \frac{1 \cdot 3 \cdots (2\nu + 1)\sqrt{\pi}}{2^{\nu}} - \beta(\nu - 1)!\right].$$
$$(50.18)$$

Substituting these values of V_{k-1} and K_{jk} into Eqs. (47.6)-(47.13) and using the expressions (47.5), (49.9), (49.12), and (50.3), we obtain in the linear approximation with respect to α_2 and β:

$$\sigma = \frac{16\pi e^2 m^*}{3h^3} k_B T \frac{e^{\tilde{\mu}}}{A}\left(1 - \frac{F' \sqrt{\pi}}{A} e^{-\frac{\theta}{T}} \sqrt{\frac{\theta}{T}}\right) ; \qquad (50.19)$$

$$u_1 = \frac{2\sqrt{2}\,e}{3\sqrt{\pi k_B T m^* A}}\left(1 - \frac{F' \sqrt{\pi}}{A} e^{-\frac{\theta}{T}} \sqrt{\frac{\theta}{T}}\right) ; \qquad (50.20)$$

$$\varkappa = \frac{4\sqrt{2}\,k_B^{3/2} \sqrt{T}\,n}{3\sqrt{\pi}\,A}\left(1 - \frac{29}{4} \cdot \frac{F' \sqrt{\pi}}{A} e^{-\frac{\theta}{T}} \sqrt{\frac{\theta}{T}}\right) ; \qquad (50.21)$$

$$\sigma_T = -\frac{k_B T}{e} \cdot \frac{d}{dt}\left(\frac{F'\sqrt{\pi}}{A}e^{-\frac{\theta}{T}}\sqrt{\frac{\theta}{T}} + \tilde{\mu}\right); \tag{50.22}$$

$$\alpha_T = \frac{k_B}{e}\left|\tilde{\mu} + \frac{F'\sqrt{\pi}}{A}e^{-\frac{\theta}{T}}\sqrt{\frac{\theta}{T}}\right|_1^2; \tag{50.23}$$

$$\sigma_H = \sigma_0\left[1 - \left(1 - \frac{\pi}{4}\right)\frac{e^2 H^2}{2k_B T m^* c^2 A^2}\right]; \tag{50.24}$$

$$R_i = \frac{3\pi}{8enc}\left[1 + \left(\frac{3\pi}{2} - 2\right)\frac{F'}{A\sqrt{\pi}}e^{-\frac{\theta}{T}}\sqrt{\frac{\theta}{T}} - \frac{\pi e^2 H^2}{8k_B T m^* c^2 A^2}\right]; \tag{50.25}$$

$$Q_i = \frac{\sqrt{2\pi k_B T}}{8T\sqrt{m^* A}}\left[1 - \left(\frac{8}{\sqrt{\pi}} - \sqrt{\pi}\right)\frac{F'}{A}e^{-\frac{\theta}{T}}\sqrt{\frac{\theta}{T}} - \left(3 + \frac{\pi}{4}\right)\frac{e^2 H^2}{2k_B T m^* c^2 A^2}\right]. \tag{50.26}$$

At low temperatures, true liquids do not exist and, therefore, the formulas (50.19)-(50.26) give the explicit temperature dependences for the transport coefficients of amorphous or glassy substances for which A and F' are independent of temperature.

All the calculations carried out in the present chapter are valid also for metals and semiconductors with p-type conduction, but in that case the signs in the expressions for σ_T, α_T, R_i, and Q_i should be reversed.

Chapter IX

LOCAL AND IMPURITY LEVELS
IN AMORPHOUS SEMICONDUCTORS

§51. Local States in the One-Dimensional Model
of a Liquid [1]

In an ideal periodic lattice, the electrons are represented by Bloch waves propagated across the whole crystal. In Chaps. IV and V, the electron wave functions for liquid and amorphous substances were represented in the form of linear combinations of Bloch's functions. These linear combinations can be used to form local states, as well as wave packets propagated across the whole sample, analogous to the well-known Wannier functions. Thus the departure from long-range order may give rise to local electron states.

This problem was first investigated by Roberts and Makinson [1] using a one-dimensional model of a liquid. Although the results obtained with the one-dimensional model (Chap. V) cannot be extended in toto to real substances, these calculations are of interest since they prove, in principle, the possibility of the existence of local states in a liquid.

The model used by Roberts and Makinson was the same as that used for the numerical calculations referred to in § 28: δ-type potential wells distributed at random. Schrödinger's equation has the form:

$$\psi'' + [k^2 + 2k_0\delta(x - x_n)]\psi = 0; \qquad (51.1)$$

here the energy is $E = k^2$, the units are selected so that $\hbar^2/2m = 1$, and $2k_0$ is the size of the δ well.

In the absence of δ wells, $\psi = A \cos k\xi$. The effect of a δ well is obtained by integrating Eq. (51.1) over a short interval near x_n. The derivative of ψ varies from ψ'_- to $\psi'_+ = \psi'_- - 2k_0\psi(x_n)$, and the amplitude variation is given by

$$\frac{A'^2}{A^2}=1+\frac{\frac{4k_0}{k}\left(\frac{k_0}{k}-\frac{\psi'_-}{k\psi}\right)}{1+\left(\frac{\psi'_-}{k\psi}\right)^2}. \qquad (51.2)$$

It is evident from Fig. 22 that the change in the slope of the $\psi(\xi)$ curve makes the half-wavelength, i.e., the distance between two nodes of ψ, approach the distance between the δ wells, and if the latter are sufficiently large, the number of nodes is equal to the number of δ wells and independent of the energy; since the number of nodes is equal to the number of states, this means that in a given energy interval no new states appear, and a forbidden band is formed (§ 28 and Figs. 18 and 19).

To investigate the conditions for the appearance of a forbidden band, it is convenient to introduce the phase ψ_n of the function $\psi(x_n)$, assuming that

$$\tan\varphi_n=\lim_{x\to x_n}\left(\frac{-\psi'}{k\psi}\right),\quad -\frac{\pi}{2}\leqslant\varphi_n\leqslant\frac{\pi}{2}. \qquad (51.3)$$

From Eq. (51.1), it is easily found that

$$\varphi_{n+1}-\varphi_n=\tan^{-1}\left(\tan\varphi_n+\frac{2k_0}{k}\right)-\varphi_n+ku_n-\pi,$$

$$u_n=x_{n+1}-x_n. \qquad (51.4)$$

In a periodic chain, all u_n are equal to u and we can find the equilibrium phase b, which does not vary with n. According to Eq. (51.4), b satisfies the equation

$$\tan(b-ku)=\tan b+\frac{2k_0}{k}. \qquad (51.5)$$

Two roots of this quadratic equation for tan b are:

$$\tan b_{\pm}=-\frac{k_0}{k}\pm\frac{\sqrt{Q^2-1}}{\sin ku}, \qquad (51.6)$$

where

$$Q(k)=-\frac{k_0}{k}\sin ku+\cos ku, \qquad (51.7)$$

$b_{\pm}$ have real values for $|Q|\geqslant 1$, and the corresponding energy interval will be called the g band. For $b_-<\varphi_n<b_+$ the right-hand part of Eq. (51.4) is greater than zero, and for $-\pi/2<\varphi_n<b_-$ or $b_+<\varphi_n<\pi/2$, it is smaller than zero.

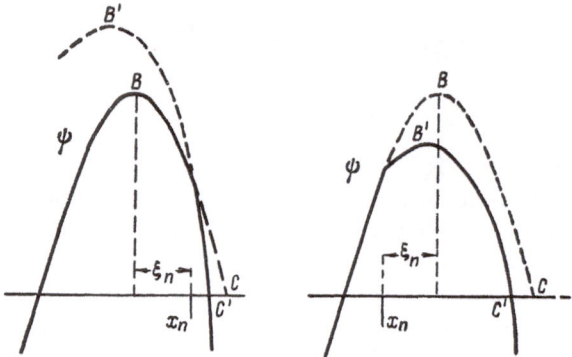

Fig. 22. Influence of a δ well on the wave function ψ. The next node approaches x_n (from C to C'). The amplitude changes (a maximum at B' instead of B).

Therefore, if $b_- < \varphi_1 < b_+$, the successive ψ_n increase monotonically tending to b_+, and if $b_+ \leq \varphi_1 < b_-$, they decrease monotonically to b_+. The function ψ has one node in each cell, i.e., the g band is forbidden. We note that the condition $|Q| > 1$ is identical with the well-known forbidden band criterion in the Kronig−Penney model.

If $-\pi/2 < \varphi_1 < b_-$, φ_n first decreases to $-\pi/2$, then φ_{m+1} "jumps" into the interval $(b_+, \pi/2)$, and further φ_n again tend to b_+. Thus in the m-th cell there is no ψ-function node; this will be a "node jump."

We shall select such φ_1 and energies E_1, E_2 (these energies are not allowed), so that

$$b_-(E_2) < \varphi_1 < b_-(E_1), \quad E_2 > E_1. \tag{51.8}$$

This is always possible because, according to Eq. (51.6), the g band occupies the interval $k_c < k < \pi/u$ and

$$b_-(k_c) = \tan^{-1} -\frac{k_0}{k_c}, \quad b_-\left(\frac{\pi}{u}\right) = -\frac{\pi}{2}. \tag{51.9}$$

However, a node jump then occurs at E_1 but not at E_2, i.e., at E_2 ψ has an extra node and there is an allowed energy level in the g band between E_1 and E_2. There is only one such level for a long chain and, therefore, it is of little importance.

These considerations acquire physical meaning for a partly disordered chain. We shall first consider a defect in a chain in the form of a truncated cell $u_j = u(1-\gamma)$, $0 < \gamma < 1/2$; $u_{n \neq j} = u$. If j is sufficiently large, then on the basis of the above treatment $\varphi_j = b_+$, and according to Eqs. (51.4) and (51.5):

$$\varphi_{j+1} = b_+ - ku\gamma. \tag{51.10}$$

If $ku\gamma > b_+ - b_-$, φ_{j+1} lies in the interval $(-\pi/2, b)$ and consequently, a node jump occurs later. Thus the presence or absence of a node jump is determined by

$$D(k) \equiv ku\gamma - [b_+(k) - b_-(k)] \gtrless 0. \tag{51.11}$$

Since $D(k_c) = k_c u\gamma > 0$, and $D(\pi/u) = \pi(\gamma - 1) < 0$, a defect gives rise to an allowed level in the g band, the level being found from $D(k) = 0$. For the wave function of this level, we have, according to Eqs. (51.10) and (51.11), $\varphi_{j+1} = b$; away from this point we again have a periodic chain and, therefore, when $n > j$ the phase φ_n becomes equal to b_- on the basis of Eqs. (51.4) and (51.5). From Eqs. (51.2) and (51.3), we find that

$$\frac{A_{n+1}^2}{A_n^2} = 1 + \frac{4k_0}{k} \cos^2 \varphi_n \left(\tan\varphi_n + \frac{k_0}{k} \right), \tag{51.12}$$

and from Eq. (51.6) we find that $-k_0/k > \tan b_- > -\infty$. Consequently, if the phase is b_-, $A_{n+1}^2/A_n^2 < 1$ and the wave function of the defect decreases exponentially from x_{j+1} to the end of the chain. On the other hand, since $-k_0/k < \tan b_+ < \infty$, the wave function amplitude rises exponentially from $x = 0$ to $x = x_j$. Thus an electron is found to be localized near a defect in the j-th cell.

We shall now apply these considerations to a liquid chain for which u_n obeys the distribution $p(u)$, cut off so that $u_{min} \leq u \leq u_{max}$ and

$$|u_{min, max} - u| \ll a, \quad ku_{max} < \pi.$$

We shall introduce b_+ and b_-, corresponding to the average distance between δ wells, as well as b'_+ and b'_-, corresponding to the minimum distance u_{min} and b''_+ and b''_- which correspond to the maximum distance u_{max}. Obviously, $b''_- < b_- < b'_- < b'_+ < b_+ < b''_+$. If φ_1 lies in the interval (b'_+, b''_+), the succeeding φ_n cannot lie outside this interval and there is an absolutely forbidden band; the conditions for this to happen are

$$2 \tan^{-1} \frac{k_0}{k} > \pi - ku_{min}, \quad ku_{max} < \pi. \tag{51.13}$$

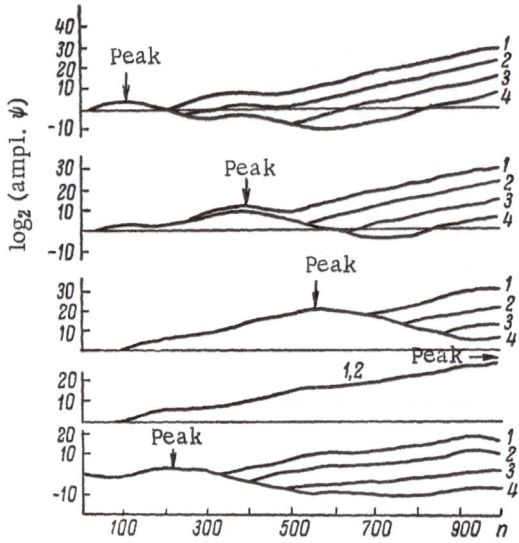

Fig. 23. Double logarithm of the amplitude
ψ near 5 eigenvalues, as a function of the
cell number. Curves 1-4 give successive approximations of the eigenvalue.

However, if the right-hand part of Eq. (51.4) is negative for sufficiently
small values of u_n and all φ_n, b''_- and b''_+ and the absolutely forbidden band
do not exist. The change in the phase for $\varphi = b_+$, b_- is on the average zero.
The region near b_+ is stable, so that, on the average, φ approaches b_+ from
both sides. However, several successive truncated cells may take the phase
over to the region $\varphi < b''_-$ and we again have a node jump.

As in the case of a single defect, it is assumed that at a certain E_1 node
jump does occur but not at E_2. Then there should be an allowed level E' in
the interval (E_1, E_2) inside the g band, which is forbidden for a periodic chain
with a period $\overline{u}$. The number of such levels is obviously determined by the
frequency of occurrence of series of closely-spaced truncated cells, and this
frequency in turn depends on the degree of disorder of the chain.

The phase φ_n of the wave function corresponding to E' lies first in the
region near b_+, then jumps to the region b_- where it remains; according to
Eq. (51.12) the average amplitude first increases and then decreases with x.
It means that an electron is localized in that part of the chain where the phase
φ_n jumps from b_+ to b_-.

Thus, in the one-dimensional model of a liquid all the electron levels are local levels inside a band which is forbidden for a crystal with the same average distance between atoms.

This result was confirmed also by numerical calculations using an electronic computer [1]. The distribution function of the distances between atoms was assumed to be:

$$\left.\begin{array}{l} p\,(u) = \dfrac{3\sqrt{5}}{20\varepsilon}\left[1 - \dfrac{(u-1)^2}{5\varepsilon^2}\right] \\[2mm] \bar{u} = 1, \quad u_{\max} = 1 + \sqrt{5}\varepsilon, \quad u_{\min} = 1 - \sqrt{5}\varepsilon. \end{array}\right\} \qquad (51.14)$$

The function of the phase distribution $w(\varphi)$ was calculated using the functional equation

$$w\,(\varphi)\,d\varphi = \sum_j p_j w\left[S_j^{-1}(\varphi)\right] dS_j^{-1}(\varphi), \qquad (51.15)$$

where

$$S_j^{-1}(x) = \tan^{-1}\left[\tan\,(x - ku_j) - \frac{2k_0}{k}\right];$$

p_j are the probabilities that $u = u_j$. The curve obtained for $w(\varphi)$ had a sharp maximum near the stable value b_+ in agreement with our discussion above.

Next, Schrödinger's equation (51.1) for a chain of 1000 cells was solved for various energies. The functions ψ and ψ' were calculated for each x_n. Curves of the amplitude of ψ were plotted as a function of n (Fig. 23) for 5 eigenvalues; curves 1-4 give successive approximations to an eigenvalue. All the curves in this figure have fairly broad peaks. An electron is found to be localized in an interval of the order of a hundred cells.

In the present discussion, we have omitted some of the finer points and proofs demonstrating that our treatment is rigorous; for the details, the reader is referred to the original work [1].

§52. Method of Calculating Local Levels in Many-Atom Systems [2]

The calculations given in the preceding section cannot be extended to the three-dimensional case. On the other hand, the effective mass method (§ 16) for calculating impurity levels in crystals is not convenient for our purpose because it does not distinguish between crystals and amorphous bodies when the same defect occurs in both cases. Moreover, the application of the effective mass method to amorphous substances is not, strictly speaking, justified because this method is associated with Bloch's waves.

To investigate the local states in liquids, we shall use a method which was validated in a general way by Lifshits [3] and applied by Koster and Slater [4] to the calculation of impurity levels in a crystal, and subsequently used by Koutetskii and Fingerland [5] in the study of the electron spectrum of an arbitrary heterogeneous system.

We shall assume that we know the wave functions ψ_i and the energies E_i for a system without a local defect in the one-electron approximation: these functions and energies satisfy Schrödinger's equation

$$\hat{H}_0\psi_i = E_i\psi_i. \tag{52.1}$$

In an amorphous substance, ψ_i represent linear combinations (30.11) of Bloch's functions (30.10).

By employing unitary transformations, we may change from the functions ψ_i to the functions φ_r localized in very restricted regions; these functions may be atomic functions, valence bond functions, Wannier functions, etc. We can also change from the functions φ_r back to the functions ψ_i. If we restrict ourselves to a single band, the appropriate unitary transformations are:

$$\varphi_r = \sum_{i=1}^{G} c_{ir}^*\psi_i, \quad \psi_i = \sum_{r=1}^{G} c_{ir}\varphi_r, \tag{52.2}$$

where G is the number of atoms or unit cells in the region considered.

The state of an electron in a system perturbed by an impurity or by some other defect is described by the equation

$$(\hat{H}_0 + V)\Phi = E\Phi, \tag{52.3}$$

where V is the additional potential of a defect. The solution of Eq. (52.3) may be found in the form of a linear combination of the localized functions

$$\Phi = \sum_{k=1}^{G} a_k\varphi_k. \tag{52.4}$$

For the coefficients a_k, we obtain equations which may be reduced to:

$$\sum_{k=1}^{G} a_k(\varepsilon_{ki} - E\delta_{ik} + V_{ik}) = 0 \quad (k = 1, 2, \ldots, G), \qquad (52.5)$$

where

$$\varepsilon_{ki} = \sum_{r=1}^{G} c_{rk}^* E_r c_{ri}, \quad V_{ik} = \int \varphi_i^* V \varphi_k d\mathbf{r}. \qquad (52.6)$$

To reduce the number of equations, we multiply Eq. (52.5) by c_{pi}^*, take the sum of i terms, and use the orthonormalization relationship

$$\sum_{i=1}^{G} c_{pi}^* c_{ri} = \delta_{pr}. \qquad (52.7)$$

Then, assuming that $E \neq E_p$, i.e., that the energy level lies outside the energy band of the unperturbed system, we divide these equations by $(E - E_p)$, multiply by c_{ps}, take the sum of p terms and finally obtain

$$a_s = \sum_{i=1}^{G} \sum_{k=1}^{G} L_{si} V_{ik} a_k \quad (s = 1, 2, \ldots, G), \qquad (52.8)$$

where

$$L_{si} = \sum_{p=1}^{G} \frac{c_{pi}^* c_{ps}}{E - E_p}. \qquad (52.9)$$

The equations (52.8) can be found directly from Eq. (19) in [5] if we assume that $P^\alpha = N^\alpha = G$ and omit the summation with respect to β and the subscripts α and β.

In the case of impurity atoms and similar defects, the potential is localized in a very small region of space and, therefore, due to the localization of the functions φ_i, the matrix elements V_{ik} are not equal to zero only for several values of i and k. If the potential V acts only on z sites, then instead of Eq. (52.8) we obtain

$$a_s = \sum_{i=1}^{z} \sum_{k=1}^{z} L_{si} V_{ik} a_k \quad \text{for} \quad s = 1, 2, \ldots, z. \qquad (52.10)$$

The above system of equations describes local states whose energies are found by equating to zero the determinant:

$$\Delta = \left| \sum_{i=1}^{z} L_{si} V_{ik} - \delta_{sk} \right| = 0. \qquad (52.11)$$

For crystals

$$c_{ps} = \frac{1}{\sqrt{G}}\, e^{ik_p \cdot r_s}, \quad c_{pi}^{*} = \frac{1}{\sqrt{G}}\, e^{-ik_p \cdot r_s},$$

and Eq. (52.11) reduces to Eq. (41) of [4]; to calculate L_{si} it is necessary to change from summation to integration in the **k** space over a volume Ω_k of a Brillouin zone:

$$L_{si} = \frac{1}{\Omega_k} \int \frac{c_{pi}^{*} c_{pk}\, d\Omega_k}{E - E'(k)}. \qquad (52.12)$$

Noting that in the calculation of L_{si} only the density of states on the E_p energy scale is important, we can represent the sum of Eq. (52.9) for a general system of atoms by the integral

$$L_{si} = \int_{0}^{E_m} \frac{O_{si}(E')\, dE'}{E - E'}, \qquad (52.13)$$

where

$$Q_{si}(E')\, dE' = \sum_{dE'} c_{pi}^{*} c_{ps}. \qquad (52.14)$$

Here the sum is taken over all states corresponding to the energy E_p in the interval from E' to E' + dE'. The energies are taken from the lower edge of the forbidden band, and E_m is the maximum energy of the band. For a crystal $Gc_{pi}^{*}c_{pi} = 1$ and Q_{ii} is equal to the density of states on the energy scale.

In other cases (for an amorphous substance, and for $s \neq i$), Q_{si} differs from the density of states but the behavior of Q_{si} is in general similar to that of an energy band.

In Chaps. IV and V, we established that the density of states in an energy band of an amorphous substance differs from the energy of states in an energy band of a crystal only by the fact that the band edges in an amorphous sub-stance are smeared out and there is a band "tail" whose effective length is of the order of ε^2 (in the three-dimensional case). Since ε is a small quantity, the contribution of this "tail" to the integral of (52.13) is not too important.

Had we not been especially interested in the difference between L_{si} for a crystal and a liquid, we could have calculated L_{si} for an amorphous substance using Eq. (52.12) for a crystal which has the same short-range order and the same average distance between atoms as the amorphous substance. The difference between L_{si} for a crystal and for an amorphous substance is mainly due to the influence of the band "tail".

The quantity V_{ik} is calculated in exactly the same way for a crystal and an amorphous substance. When the values of V_{ik} and L_{si} are known, the energy level E of the local state is easily found from Eq. (52.11).

§53. Local Fluctuation Levels in Amorphous Semiconductors [6]

The results obtained in § 51 for the one-dimensional model cannot be applied directly to a real three-dimensional liquid. As shown in Chap. V, the one- and three-dimensional models of a liquid give completely different results. We have mentioned also that in the one-dimensional case any potential well, no matter how shallow, gives rise to a local level, while in the three-dimensional case a local level appears only if the well is sufficiently large, i.e., if the product of its depth and the square of its radius is sufficiently large.

Nevertheless, we may assume that in a three-dimensional liquid local levels should exist even in the absence of impurities. This is because large short-range order fluctuations, although rare, do occur in a liquid and they give rise to local disturbances sufficiently strong to give rise to local fluctuation states.

The curve representing the interaction between atoms is asymmetric with respect to the point of minimum energy: on reducing the distance between atoms the energy increases more strongly than when the distance is increased. Due to this, the function which gives the distribution of distances between neighboring atoms in a liquid should also be asymmetric with respect to the average distance. On the short distance side, this distribution function is cut off but the number of small departures on the short side is large; the maximum of the distribution function may be displaced toward short distances from the average value. Conversely, on the long-distance side, the distribution function may have a long tail, but the number of deviations in the direction of larger distances is relatively small. Consequently, local aggregates of atoms will not cause the splitting of local levels from an energy band (which one might expect in the case of the relatively rare but nevertheless considerable disturbances of atomic ordering) and will not give rise to holes, characteristic of liquids and responsible for a strong perturbing potential.

To investigate local fluctuation levels, we shall use the method presented in the preceding section.

We shall consider a large fluctuation separately and we shall regard it as a local perturbation giving rise to a potential V. The unperturbed system is an amorphous substance with small short-range order fluctuations. The energy band of this system differs little in its density of states from a crystal band and therefore the quantities L_{si} can be calculated approximately using Eq. (52.12). The local level energy will be found from Eq. (52.11).

We shall consider Eq. (52.11) separately for two extreme types of semiconductors: ionic and covalent.

The valence band of ionic semiconductors is usually formed from levels of negative ions, and the conduction band is formed from free levels of positive ions. A conduction electron may be localized near one of the positive ions if an additional potential well is formed as a result of an increase of the distance between the nearest negative ions. If the perturbation is localized in one cell, we must assume that $z = 1$ and that approximately $G|c_p|^2 = 1$, as in the case of crystals. Substituting Eq. (52.9), we find that Eq. (52.11) becomes:

$$\frac{1}{V_{11}} = \frac{1}{\Omega_k} \int \frac{d\Omega_k}{E - E'(\mathbf{k})}. \qquad (53.1)$$

This integral is similar to the integral of Eq. (32.5), obtained in Chap. V, and is calculated as before.

For $E < 0$ (E is taken from the lower edge of the conduction band), the integral (53.1) may be calculated in the effective mass approximation, assuming that $d\Omega_k = 4\pi k^2 dk = 4\pi \sqrt{2E'} m^{*3/2} \frac{dE}{\hbar^3}$ and omitting the small contribution of the upper part of the band. Consequently, we have

$$\frac{1}{V_{11}} = \frac{8\pi m^{*3/2} \sqrt{2}}{\Omega_k \hbar^3} \left(\sqrt{-E} \tan^{-1} \sqrt{\frac{E_1}{-E}} - \sqrt{E_1} \right), \qquad (53.2)$$

where E_1 lies in the middle of the forbidden band.

This equation has a negative solution ($E < 0$) if

$$|V_{11}| > \frac{\Omega_k \hbar^3}{8\pi m^{*3/2} \sqrt{2E_1}}. \qquad (53.3)$$

The orders of magnitude are $\Omega_k \approx \pi^3/a^3$, $E_1 \approx \hbar^2/2m^*a^2$ (a is the cell dimension). In the first approximation, V_{11} is equal to the perturbed potential at the center of a positive ion, calculated using the point-ion model:

$$V_{11} = e^2 \frac{e^*}{e} \left(\sum_{\nu=1}^{z_1} \frac{1}{r_\nu} - \frac{z_1}{z_0} \right) \simeq \frac{e^2}{r_0} \cdot \frac{e^*}{e} \sum_{\nu=1}^{z_1} \frac{\Delta r_\nu}{r_0} ; \qquad (53.4)$$

here e^* is the effective charge of an ion, which takes into account the degree of ionic binding, the overlap of electron shells of neighboring ions, and the ionic polarizability; r_0 is the average distance between neighboring ions; Δr_ν represents fluctuations of this distance; and z_1 is the number of ions in the first coordination shell.

The inequality (53.3) gives the condition for the appearance of a local level:

$$\sum_{\nu=1}^{z_1} \frac{\Delta r_\nu}{r_0} > b \equiv \frac{a_0}{a} \cdot \frac{r_0}{a} \cdot \frac{e}{e^*} \cdot \frac{m}{m^*} ; \qquad (53.5)$$

$a_0 = \hbar^2/me^2$ is the Bohr radius.

The number of fluctuation levels can be found by assuming that the quantities $t_\nu = \Delta r_\nu/r_0$ in the positive region obey the normal distribution with a dispersion ε, so that $\overline{t^2}_\nu = \varepsilon^2$, where ε is the degree of departure from short-range order. The sum of z_1 random quantities $t = \sum_{\nu=1}^{z_1} t_\nu$, obeys the normal distribution with a dispersion $\varepsilon\sqrt{z_1}$

$$f(t) = \frac{1}{\varepsilon \sqrt{2\pi z_1}} e^{-\frac{t^2}{2\varepsilon^2 z_1}} , \qquad (53.6)$$

and the local level density is given by

$$N = N_0 \frac{1}{\varepsilon \sqrt{2\pi z_1}} \int_b^\infty e^{-\frac{t^2}{2\varepsilon^2 z_1}} dt \simeq N_0 \frac{\varepsilon \sqrt{z_1}}{b \sqrt{2\pi}} e^{-\frac{b^2}{2\varepsilon^2 z_1}} , \qquad (53.7)$$

where N_0 is the number of positive ions in 1 cm^3. The local level density depends strongly on the quantity b, which may vary from substance to substance. We note that $r_0 \approx a$, $a_0 < a$; usually $e^* < e$ and $m^* < m$; in our example we shall assume that $b = 1$. If $N_0 = 10^{22}$ cm^{-3}, $z_1 = 6$ and $\varepsilon = 0.1$, then $N = 2.5 \cdot 10^{17}$ cm^{-3}.

These levels are split from the conduction band and are acceptors or electron traps. In exactly the same way, we can consider a negative ion around which positive ions are pushed further apart and this leads to the appearance of a donor level or a hole trap.

As mentioned in § 5, the necessary condition for the formation of glass in a semiconducting system is the presence of covalent bonds in the melt with the predominance of a chain structure.

Covalent semiconductors may be treated by the equivalent orbital method in the zeroth-order approximation, regarding each valence bond as a diatomic molecule. The energy of an electron in a molecule consisting of two identical atoms a and b is, in the one-electron approximation [7],

$$E \simeq E_a - U_{aa} \mp U_{ab},\qquad (53.8)$$

where

$$U_{aa} = -\int \varphi_a^* U_a \varphi_a dr; \quad U_{ab} = -\int \varphi_b^* U_a \varphi_a dr; \qquad (53.9)$$

E_a is the energy of a valence electron of the a-th atom; φ_a and φ_b are the valence wave functions of the atoms a and b; U_a is the perturbing potential due to the b-th atom, acting on the electron of the a-th atom; $U_a < 0, U_{ab} > 0$. The upper sign in Eq. (53.8) represents the bonding state; the lower sign gives the antibonding state. We may assume that [8] the valence band is formed from the bonding states and the conduction band in the antibonding state.

In the case of a typical covalent bonding $|U_{aa}| \ll U_{ab}$ (see p. 143 in [7]), as U_{ab} decreases the distance r between atoms increases. Thus, a valence bond stretched by fluctuations represents a potential well for an electron in the conduction band and a potential hill for an electron in the valence band; here, of course, we mean the pseudopotential to which the Pauli principle applies: it is the deformation potential introduced in [9].

The foregoing considerations are in agreement with the experimental observation that the forbidden band of covalent semiconductors narrows with increasing volume.

If we assume that the order of magnitude is given by $\Delta U_{ab}/U^0_{ab} = -\Delta r/r$, the perturbing potential for a conduction electron is

$$V \simeq -U^0_{ab} \frac{\Delta r}{r_0}. \qquad (53.10)$$

Later calculations will show that a conduction electron is more likely to be localized not at one bond but at z neighboring bonds if they are stretched at the same time. In order to carry out calculations using Eq. (52.11) for $z > 1$, we shall assume that $Gc^*{}_s c_p = 1$, so that all L_{si} are identical. Next, we replace $\sum_{i=1}^{z} V_{iq}$ by its q average:

$$\frac{1}{z} \sum_{q=1}^{z} \sum_{i=1}^{z} V_{iq} = \frac{\varphi}{z} \sum_{q=1}^{z} V_{qq} = \varphi \bar{V}_{qq}. \qquad (53.11)$$

The coefficient φ depends on the degree of overlap of the bonding functions φ_i. The determinant (52.11) simplified in this way may be reduced to the diagonal form by subtracting certain rows and columns. After some simple transformations, Eq. (52.11) becomes:

$$\frac{1}{\bar{V}_{qq}} = \frac{z\varphi}{\Omega_k} \int \frac{d\Omega_k}{E - E'(k)}, \qquad (53.12)$$

which is similar to Eq. (53.1).

According to Eq. (53.10),

$$\bar{V}_{qq} = -\frac{U^0_{ab}}{z} \sum_{q=1}^{z} \frac{\Delta r_q}{r_0}. \qquad (53.13)$$

We note that $2U^0_{ab}$ is approximately equal to the distance between the middle points of the valence and conduction bands, i.e.,

$$U^0_{ab} \simeq E_1 + \frac{E_g}{2} \approx \frac{\hbar^2}{2m^* a^2}.$$

Having carried out the same calculations as in the derivation of Eq. (53.5), we obtain the condition for the existence of a local level:

$$\sum_{q=1}^{z} \frac{\Delta r_q}{r_0} > \frac{2}{\varphi} \qquad (53.14)$$

and the expression for the local level density:

$$N = N_0 \frac{1}{\varepsilon \sqrt{2\pi z}} \int_{\frac{2}{\varphi}}^{\infty} e^{-\frac{t^2}{2\varepsilon^2 z}} \, dt \simeq N_0 \frac{\varphi \varepsilon \sqrt{z}}{2\sqrt{2}\,\pi} e^{-\frac{2}{\varepsilon^2 z \varphi^2}}, \qquad (53.15)$$

where N_0 is the number of valence bonds in 1 cm^3.

N rises sharply with z, but at large values of z an electron ceases to be localized due to the overlap of the wave functions of various fluctuation potential wells. An electron remains localized only if

$$Nz \ll N_0. \qquad (53.16)$$

If $\varepsilon = 0.1$, $\varphi \simeq 2$, the condition (53.16) is satisfied when $z \lesssim 10$; for example, if $z = 10$ we have $N = 8.5 \cdot 10^{18}$.

Similarly, we can find donor levels split from the valence band, but because the bonding function localization is greater, the value of φ is smaller. For example, if we assume that $\varphi = 1$, then $N = 2 \cdot 10^{12}$ for $\varepsilon = 1$ and $z = 10$, i.e., the local level density is much lower than the acceptor concentration.

All these numerical estimates should be regarded only as examples. The fluctuation level densities may differ by several orders of magnitude from substance to substance.

The local levels considered here are not impurity levels because they appear in a system of identical atoms. The fluctuation levels are similar in their origin to crystalline defect levels of vacancy or interstitial-atom type, but they differ quite considerably from the latter in other respects.

First, crystalline defect levels form discrete spectra while fluctuation levels form continuous spectra. According to Eq. (53.1) or (53.12), the stronger the fluctuations the deeper the level position in the forbidden band. If the dimensions of these fluctuations have a Gaussian distribution, the levels have the same distribution and their number should decrease rapidly away from the upper edge of the forbidden band.

Secondly, fluctuation levels in liquids exist for a short time. The lifetime of a local configuration of atoms is, according to Eq. (23.1):

$$\tau = \tau_0 e^{\frac{W}{k_B T}}; \qquad (53.17)$$

here τ_0 is the period of atomic vibrations, which is approximately 10^{-13} sec; W is the activation energy for the rearrangement of the atoms, which can be found from the temperature dependence of the viscosity [10].

According to the viscosity estimates, $\tau = 10^{-11} - 10^{-12}$ sec for the majority of molten salts; we can expect the same value for molten semiconducting oxides and sulfides, which retain their short-range order and remain semiconductors after melting (§ 2).

The lifetime τ should be compared with τ_e, the residence time of an electron at a local level, which usually lies between 10^{-9} and 10^{-11} sec. For a finite lifetime of a local level, the average density of electrons at this level decreases in the ratio $\tau/(\tau_e + \tau)$ and if $\tau \ll \tau_e$ the local level does not survive sufficiently long to capture an electron; thus it plays no role in electron capture (this is probably true of the majority of liquid semiconductors).

The situation is different in glassy semiconductors. Here, the fluctuations in the distributions of the atoms are frozen and $\tau \gg \tau_e$, i.e., fluctuation levels have the same role as local levels, acting as electron and hole traps.

§ 54. Impurity Levels in Amorphous Semiconductors [2]

We shall now consider the characteristic features of the behavior of normal donors or acceptors in amorphous semiconductors. The first attempt to investigate theoretically this problem was made by Fisher [11] but his results were not satisfactory.

Fisher concluded that the action of a field fluctuating in space destroys the bound state of an electron at an impurity atom, in contrast to the periodic field of a crystal which does not affect the bound state. However, such a conclusion is not justified. The local state itself is a departure from periodicity and therefore without solving Schrödinger's equation it is not obvious that a nonperiodic potential of the surrounding medium can have an effect which is basically different from that of a periodic potential.

On the other hand (§ 51 and 53), the nonperiodicity of a potential by itself leads to the appearance of local states.

We shall investigate the behavior of impurity levels in an amorphous substance more rigorously than did Fisher. We shall assume that an impurity atom in an amorphous substance gives rise to approximately the same potential as in a crystal but that an impurity level may occupy a different position than in a crystal because of the fluctuation potential background which is not strictly periodic. In this case, the calculations should be carried out using the method presented in § 52.

To carry out a quantitative study we shall consider only a very localized defect when $z = 1$ and Eq. (52.11) gives

$$L_{11} = \frac{1}{V_{11}}. \tag{54.1}$$

In order not to lose the difference between crystals and amorphous substances, L_{11} should be calculated using Eq. (52.13).

The influence of the nature of the unperturbed system on the impurity level is given solely by the dependence $Q(E')$. The quantity Q represents roughly the behavior of an energy band. On transition to the amorphous state, the allowed energy bands broaden slightly, their edges become diffuse, and tails appear.

Since $E < 0$ if $V_{11} < 0$, we find from Eqs. (54.1) and (52.13) that

$$E = V_{11} \int_0^{E_m} \frac{Q(E')\, dE'}{1 + \dfrac{E'}{|E|}}. \qquad (54.2)$$

An analysis of the above equation shows that the band broadening and the spreading of the band edges reduces the absolute value of E. For example, for a very narrow band when $E_m \ll |E|$ the integral in Eq. (54.2) becomes unity due to the normalization of the coefficients c_{p1} and the fact that $E = V_{11}$. When the band broadens, the denominator in Eq. (54.2) increases and $|E|$ decreases, i.e., $|E| < |V_{11}|$.

On transition to the amorphous state, the shallow impurity levels may merge with the conduction band tail, but, contrary to the conclusion of Fisher [11], this effect is very weak since the band broadening due to this transition is slight.

We shall now assume that impurity atoms in an amorphous state are in approximately the same positions as in a crystal but they give rise to a considerable rearrangement of the neighboring atoms so that on the whole a defect gives rise to a different perturbation potential than in a crystal.

Due to the absence of strict order in the distribution of the atoms, the displacement of the atoms near a defect in an amorphous substance is considerably greater than in a crystal. In a crystal, the lattice is elastically deformed near a defect but in an amorphous substance local plastic deformation may occur. This hypothesis applies to the interstitial atoms, vacancies, and substitution impurities, but the mechanism of the level displacement is different in these three cases.

For an interstitial atom, we may use the effective mass method or the tight binding method described above.

Using the effective mass method to calculate the levels of hydrogen and lithium atoms in germanium, Reiss [12] assumed that an atom in an interstice may be represented by a spherical cavity in a dielectric continuum, the cavity enclosing the atomic core. The potential of the atom is in this case

$$U = \frac{e}{r} - \frac{e}{r}\left(1 - \frac{1}{\varepsilon}\right)\ (r < a), \quad U = \frac{e}{\varepsilon r}\ (r > a), \qquad (54.3)$$

where ε is the permittivity; a is the radius of the cavity determined by the dimensions of the interstice.

Variational calculations show that with the increase of a from 0 to ∞ the depth of the level $|E|$ increases by a factor of ε^2, and that a considerable rise begins when $a > r_s$, where r_s is the radius of the valence orbital of the atom in the cavity.

For hydrogen atoms in germanium, $a > r_s$ and therefore the impurity levels are deep and do not act as donors. On the other hand, $a < r_s$ for lithium atoms whose levels are shallow so that lithium acts as a donor. These conclusions are in agreement with the experimental observations.

The displacement of the surrounding atoms in the effective mass method may be described by a change in the cavity dimensions.

In a crystal, the cavity radius is determined in the first approximation by the dimensions of a natural interstice, whereas in amorphous substances the cavity radius is determined by the dimensions of an impurity atom because the neighboring atoms may be pushed apart. It follows that the cavity radius in amorphous substances is much larger than that in crystals. Therefore, according to Reiss, an impurity atom level in an amorphous substance is considerably deeper than that in a crystal.

The same result is obtained by the method of tightly bound electrons. The displacement of neighboring atoms changes the perturbation potential U. Every electron outside the filled band is repelled from filled atomic shells and valence bonds. The repulsion energy has been included in the detailed calculations of local levels in crystal phosphors carried out by Williams [13] and Potekhina [14]. According to Potekhina, the repulsion potential decreases rapidly with distance between the impurity atom and its neighbors. Therefore, in an amorphous substance where neighboring atoms are displaced, the repulsion energy is considerably lower than in crystals; this reduces the depth of the impurity electron level. This reduction is considerably greater than the upward shift of the level for fixed U obtained above.

If we have a vacancy, the repulsion energy near it is less than near filled sites and therefore a vacancy is a potential well for an electron from the conduction band and acts as an acceptor whose level lies well below the bottom of the conduction band. However, a vacancy of normal dimensions is stable only in a fully ordered crystalline lattice. In an amorphous substance, the rearrangement of the neighboring atoms compresses the cavity and reduces the potential well thereby raising the energy level.

We shall now consider substitutional impurities. If an atom A, forming z valence bonds with neighboring atoms is replaced in a predominantly covalent crystal with another atom M which has a smaller number of valence electrons (for example, $z-1$), then the M atom forms only $z-1$ valence bonds, so that if the M atom is surrounded symmetrically by z A atoms, $z-1$ bonds are not completely localized but are distributed between neighbors, giving rise to "valence structure resonance." Such a system tends to form a total of z bonds, and therefore easily captures an extra electron and acts as an acceptor with a low-lying level.

Obviously, a system with $z-1$ "free" valence bonds is not stable by itself and the symmetric positions of z neighboring atoms are retained only by the influence of the surrounding ordered lattice. In an amorphous substance, this symmetry is disturbed, as in the case of hydrogen bonds in organic molecules [15]: one of the neighboring atoms is slightly further and the remaining $z-1$ atoms form stronger fully localized bonds with the M atom. The low-lying level then disappears.

Similar considerations apply to an impurity atom X which has $z+1$ electrons, and acts as a donor in a crystal because it gives up easily its excess electron which does not take part in valence bonding. In an amorphous substance, there are two possibilities. By a rearrangement of neighboring atoms, the X atom forms valence bonds with $z+1$ atoms, or two electrons are paired in an inner shell of the X atom so that this atom forms bonds with $z-1$ atoms. In both cases, there is an unpaired electron at a high-lying donor level.

From these considerations, we can draw the following general conclusion: local states with high-lying occupied or low-lying empty levels are unstable, and on transition to the amorphous state, the atoms near defects are rearranged so that the energy spectrum is "smoothed out": the empty levels approach the free band and the occupied levels approach the filled band.

§ 55. Possible Reasons for the Absence of Impurity Conduction in Amorphous Semiconductors

The simplest assumption is that impurity atoms in amorphous substances occupy different sites than in crystals. If, for example, in a crystal an impurity atom occupies a lattice site (substitutional impurity), in an amorphous substance it may occupy an interstice—where, of course, it behaves in a different way. The probability of an impurity occupying an interstice in an amorphous substance is higher than in a crystal because of the looser structure of the amorphous substance, the absence of strict order in the distribution of the atoms, and the resultant presence of enlarged interstices.

This hypothesis cannot, however, explain the complete absence of impurity conduction in glasses. If a glass has the same chemical structure (short-range order) as a crystal, then a substitutional impurity may occupy a regular site. The numbers of impurity atoms at lattice sites and interstices are related by a statistical relationship so that the size of an impurity in an amorphous substance is important only at higher concentrations because only some of the impurity atoms are active.

The merging of impurity levels with the "tail" of a band or the disappearance of an impurity level in the fluctuation potential background, predicted by Eq. (54.2), cannot by themselves explain the experimental observations. If, for example, the donor levels lying close to the lower edge of a free band merge with this band and their electrons enter this band, we find, assuming that Bloch-type progressive waves represent the band, that the conductivity is of the metallic type, thereby contradicting the experimentally observed intrinsic conductivity. Similarly, we can deal with the merging of acceptor levels with a filled band and the appearance of p-type metallic conductivity.

There remain two possible explanations of the absence of the impurity conduction in glassy semiconductors.

1. We have shown that, due to a rearrangement of the atoms surrounding an impurity atom, the energy spectrum is smoothed out so that free levels (acceptors) approach a free band and filled levels (donors) approach a filled band. In this case, donors and acceptors, like other levels well within the forbidden band, cease to be important and the conduction becomes intrinsic.

2. The theoretically predicted (§ 53) local fluctuation levels capture electrons or holes and thus neutralize the donors and acceptors. For this purpose, there should be a sufficient number of fluctuation levels well within

the forbidden band. The density of fluctuation levels decreases away from the allowed bands in accordance with the Gaussian law. Consequently, the "tail" of this Gaussian distribution should be strong in order to neutralize shallow donors or acceptors. It is possible that the presence of long chains of atoms in glasses favors the formation of deep levels and their energy distribution need not be Gaussian but may have a maximum at some depth.

As shown in § 53, the concentration of fluctuation acceptors in covalent semiconductors is higher than is the concentration of fluctuation donors. This explains the observation that glassy semiconductors usually have n-type conduction.

The fluctuation levels in glassy semiconductors are much more important than those in liquids. This means that liquid semiconductors may exhibit impurity conduction if its absence in glasses is due to fluctuation levels. Experiments show (§ 2) that the electrical conductivity of some semiconductors changes only slightly on melting. However, this may be explained also by the fact that the conduction in a solid semiconductor becomes intrinsic at high temperatures near the melting point. To obtain information on this point, further experimental studies are needed.

So far we cannot decide between the two possible explanations of the absence of impurity conduction. The second explanation is more acceptable in the case of substitutional impurities when the rearrangement of the surrounding atoms is less likely.

There is no basic difference between the two explanations. We may assume that the first explanation means that an impurity atom becomes lodged in a region of fluctuations of the positions of other atoms or gives rise to such a fluctuation itself. Then the difference between the first and second explanations reduces simply to the question of whether an impurity atom and a fluctuation responsible for a deep level coincide in space.

Chapter X

ELECTRON STRUCTURE
OF LIQUID METALS [1, 2]

§56. Calculation Method

A theory of amorphous conductors, based mainly on the deformed coordinate method, was presented in Chaps. IV–IX. It was assumed that the
energy spectrum and wave functions of the electrons in a crystal are known,
and the changes in the energy spectrum and other properties occurring on melting were calculated. This theory applies both to metals and semiconductors
but we are more interested in semiconductors because of the importance of the
band structure of the energy spectrum. Although the various parameters occurring in this theory can, in principle, be determined if we know the spectrum
and wave functions of electrons in a crystal, in practice it is difficult to calculate these parameters. Therefore, the theory gives general and qualitative
results but not quantitative ones for actual substances.

In view of this, it is interesting to adopt a different approach, which is
not universal but is capable of giving a mathematically more rigorous actual
result for a certain group of amorphous conductors, irrespective of whether the
wave functions for the corresponding crystals are known or not. This approach
can be used in the case of liquid metals. S. F. Edwards applied the method of
weakly interacting electrons using Green's functions to a one-dimensional
model of a disordered system [1] and then to three-dimensional liquid metals [2].

The use of the method of weakly interacting electrons is not justified except perhaps for univalent metals. Therefore, one should treat with caution
the results obtained by this method particularly in respect to forbidden bands.
Nevertheless, since the presence of forbidden bands is unimportant for the majority of metal properties, this method is interesting although quite inapplicable
to semiconductors.

The calculations are essentially the same for the three-and one-dimensional models and therefore we shall quote the results for the three-dimensional

model only [2]. Schrödinger's equation for an electron moving in a field of atoms has the form:

$$-ih\frac{\partial}{\partial t}-\frac{\hbar^2}{2m}\nabla^2+\sum_n U(\mathbf{r}-\mathbf{R}_n)\,\psi=0, \qquad (56.1)$$

where $U(\mathbf{r}-\mathbf{R}_n)$ is the potential established at a point $\mathbf{r}$ by an atom located at $\mathbf{R}_n$. To determine the density of states given by Eq. (56.1), it is more convenient to solve the appropriate inhomogeneous equation for Green's functions $G_\pm$

$$\left[-i\hbar\frac{\partial}{\partial t}-\frac{\hbar^2}{2m}\nabla^2\pm i\iota+\sum_n U(\mathbf{r}-\mathbf{R}_n)\right]G_\pm(\mathbf{r},\mathbf{r}',t,t')= \qquad (56.2)$$
$$=\delta(\mathbf{r}-\mathbf{r}')\,\delta(t-t'),$$

which may be dealt with by a special approximate method. The small imaginary term $\pm i\iota$ determines whether the Green's function corresponds to an ingoing (G_+) or an outgoing (G_-) wave.

As shown in [1], the density of states, i.e., the probability of finding an electron of energy E and momentum $\mathbf{k}$, $\rho(E,\mathbf{k})$, is related to Green's functions, i.e., it is the Fourier transform of the quantity

$$\rho(\mathbf{r}-\mathbf{r}',\,t-t')=\frac{i}{2\pi}(\bar{G}_+-\bar{G}_-); \qquad (56.3)$$

here and later, a bar denotes averaging over the distribution of R_n.

Since, in the absence of external fields, the space and time are uniform, $G_\pm(\mathbf{r},\mathbf{r}',t,t')$ are functions of $\mathbf{r}-\mathbf{r}'$ and t−t'.

The average density of states on the energy scale is:

$$n(E)=\frac{\Omega}{(2\pi)^3}\int\rho(E,\mathbf{k})\,d\mathbf{k}, \qquad (56.4)$$

where Ω is the total volume occupied by the system; integration with respect to $d\mathbf{k}$ means integration over the $\mathbf{k}$ space.

To find G by the perturbation method, we use an expansion in powers of the difference between the potential energy and its average value:

$$V(\mathbf{r})=\sum_n U(\mathbf{r}-R_n)-\sum_n \overline{U(\mathbf{r}-\mathbf{R}_n)}. \qquad (56.5)$$

To simplify our treatment, we select the zero energy to be such that the second term in Eq. (56.5) vanishes. Expanding Eq. (56.2) in series of powers of V, we obtain

$$G_{\pm}(\mathbf{r}, \mathbf{r}', t, t') = G_0(\mathbf{r}, \mathbf{r}', t, t') -$$

$$- \int G_0(\mathbf{r}, \mathbf{r}'', t, t'') V(\mathbf{r}'') G_0(\mathbf{r}'', \mathbf{r}', t'', t') d\mathbf{r}'' dt'' +$$

$$+ \int G_0(\mathbf{r}, \mathbf{r}'', t, t'') V(\mathbf{r}'') G_0(\mathbf{r}'', \mathbf{r}''', t'', t''') V(\mathbf{r}''') \times$$

$$\times G_0(\mathbf{r}''', \mathbf{r}', t''', t') d\mathbf{r}'' d\mathbf{r}''' dt'' dt''' - + \ldots, \tag{56.6}$$

where the integration with respect to $d\mathbf{r}$ denotes integration over the volume; the Green's function G_0, which does not contain potential, satisfies the equation

$$\left(-i\hbar \frac{\partial}{\partial t} - \frac{\hbar^2}{2m} \nabla^2 + i\iota\right) G_0(\mathbf{r} - \mathbf{r}', t - t') = \delta(\mathbf{r} - \mathbf{r}') \delta(t - t'), \tag{56.7}$$

and its Fourier transform $G_0(\mathbf{k}, E)$ satisfies

$$\left(E - \frac{\hbar^2}{2m} k^2 + i\iota\right) G_0(\mathbf{k}, E) = 1. \tag{56.8}$$

The expansion (56.6) is known in the theory of Green's functions (see, for example [3] or [4]). For brevity, we shall omit the arguments t, t', and E of the functions G, and, in cases where this will not lead to misunderstanding, we shall also omit $\mathbf{r}$ and $\mathbf{k}$.

Kohn and Luttinger [5] have shown that we can average out all possible atomic configurations under the integral symbols in Eq. (56.6). Such a procedure would have been impossible in the initial equation, Eq. (56.1), and this is the advantage of the method proposed here.

The averaging makes the first integral in Eq. (56.6) vanish, on the basis of Eq. (56.5); all the remaining terms, which are odd with respect to V, are small and we can neglect them [6]. Thus, we obtain

$$G_+ = G_0 + \int G_0(\mathbf{r}, \mathbf{r}'') \overline{V(\mathbf{r}'') G_0(\mathbf{r}'', \mathbf{r}''') V(\mathbf{r}''')} G_0(\mathbf{r}''', \mathbf{r}') d\mathbf{r}'' d\mathbf{r}''' +$$

$$+ \int G_0 \overline{V G_0 V G_0 V G_0 V} G_0 d\mathbf{r}'' d\mathbf{r}''' d\mathbf{r}^{IV} d\mathbf{r}^V + \ldots \tag{56.9}$$

The second term on the right represents the averaging

$$\sum_{n,\,m} \overline{U(\mathbf{r}_1 - \mathbf{R}_n)\, U(\mathbf{r}_2 - \mathbf{R}_m)} =$$

$$= \frac{1}{(2\pi)^6} \int u\,(\mathbf{k})\, u\,(\mathbf{k}')\, e^{i\mathbf{k}\,\cdot\,\mathbf{R}_n + i\mathbf{k}'\,\cdot\,\mathbf{R}_m} e^{-i\mathbf{k}\,\cdot\,\mathbf{r}_1 - i\mathbf{k}'\,\cdot\,\mathbf{r}_2}\, d\mathbf{k}d\mathbf{k}', \quad (56.10)$$

where

$$u\,(\mathbf{k}) = \int U(\mathbf{r})\, e^{i\mathbf{k}\,\cdot\,\mathbf{r}} d\mathbf{r}. \qquad (56.11)$$

Since the averaged out system is uniform

$$\sum_{n,\,m} \overline{e^{i\mathbf{k}\,\cdot\,\mathbf{R}_n + i\mathbf{k}'\,\cdot\,\mathbf{R}_m}} = \frac{(2\pi)^3}{\Omega}\, \delta\,(\mathbf{k} + \mathbf{k}')\, c_2\,(\mathbf{k}); \qquad (56.12)$$

here $c_2\,(\mathbf{k})$ is a binary correlation function in the $\mathbf{k}$ space, i.e., it is the Fourier transform of the binary distribution function

$$c_2\,(\mathbf{k}) = \sum_{n,\,m} \overline{e^{i\mathbf{k}\,\cdot\,(\mathbf{R}_n - \mathbf{R}_m)}}. \qquad (56.13)$$

Thus,

$$\sum_{n,\,m} U(\mathbf{r}_1 - \mathbf{R}_n)\, U(\mathbf{r}_2 - \mathbf{R}_m) =$$

$$= \frac{1}{(2\pi)^3\,\Omega} \int |\,u\,(\mathbf{k})\,|^2\, c_2\,(\mathbf{k})\, e^{i\mathbf{k}\,\cdot\,(\mathbf{r}_1 - \mathbf{r}_2)}\, d\mathbf{k}. \qquad (56.14)$$

The Fourier transform of the second term in Eq. (56.9) has the form:

$$G_0\,(\mathbf{k})\, \Sigma_2\,(\mathbf{k})\, G_0\,(\mathbf{k}),$$

where

$$\Sigma_2\,(\mathbf{k}) = \frac{1}{\Omega} \int c_2\,(\mathbf{k} - \mathbf{k}')\, G_0\,(\mathbf{k})\,|\,u\,(\mathbf{k} - \mathbf{k}')\,|^2\, d\mathbf{k}'. \qquad (56.15)$$

Similarly, the third term in Eq. (56.6), containing V^4, may be reduced to

$$G_0\,(\mathbf{k}) \int c_4\,(\mathbf{k}',\, \mathbf{k}'',\, \mathbf{k}''',\, \mathbf{k}^{\mathrm{IV}})\, G_0\,(\mathbf{k} - \mathbf{k}')\, G_0\,(\mathbf{k} - \mathbf{k}' - \mathbf{k}'') \times$$
$$\times G_0\,(\mathbf{k} - \mathbf{k}' - \mathbf{k}'' - \mathbf{k}''')\, u\,(\mathbf{k}')\, u\,(\mathbf{k}'')\, u\,(\mathbf{k}''')\, u\,(\mathbf{k}^{\mathrm{IV}}) \times$$
$$\times \delta\,(\mathbf{k}' + \mathbf{k}'' + \mathbf{k}''' + \mathbf{k}^{\mathrm{IV}})\, d\mathbf{k}'d\mathbf{k}''d\mathbf{k}'''d\mathbf{k}^{\mathrm{IV}} G_0\,(\mathbf{k}), \qquad (56.16)$$

where c_4 is given by the equation

$$c_4\,(\mathbf{k}',\, \mathbf{k}'',\, \mathbf{k}''',\, \mathbf{k}^{\mathrm{IV}})\, \delta\,(\mathbf{k}' + \mathbf{k}'' + \mathbf{k}''' + \mathbf{k}^{\mathrm{IV}}) =$$

$$= \frac{\Omega}{(2\pi)^3} \sum_{n,\,m,\,p,\,q} e^{i(\mathbf{k}'\mathbf{R}_n + \mathbf{k}''\mathbf{R}_m + \mathbf{k}'''\mathbf{R}_p + \mathbf{k}^{IV}\mathbf{R}_q)}. \qquad (56.17)$$

If the integral in Eq. (56.16) is written in the form:

$$\sum_2(\mathbf{k})\, G_0(\mathbf{k})\, \sum_2(\mathbf{k}) + \sum_4(\mathbf{k}), \qquad (56.18)$$

the series (56.6) becomes

$$\bar{G}(k) = G_0 + G_0 \sum_2 G_0 + G_0 \sum_2 G_0 \sum_2 G_0 + G_0 \sum_4 G_0 + \cdots \qquad (56.19)$$

In some cases, Σ_4 is small compared with Σ_2, i.e., c_4 splits into a product of two c_2's; similarly c_6 splits into a product, and so on. Then the series (56.19) is approximately a geometric series and may be summed to give

$$\left[E - \frac{\hbar^2}{2m} k^2 + i\iota - \sum_2(\mathbf{k}) - \sum_4(\mathbf{k}) + \cdots \right] \bar{G} = 1. \quad (56.20)$$

§57. Calculation of the Density of States in Limiting Cases

To illustrate the theory presented in the present chapter, we shall consider two exteme cases.

Case of Complete Disorder

In this case, only the terms with $n = m$ are not equal to zero in Eq. (56.13). The number of such terms is N, so that

$$c_2(\mathbf{k}) = N. \qquad (57.1)$$

In the sum of Eq. (56.17), the terms with $n = m$, $p = q$; $n = p$ $m = q$; $n = q$, $m = p$; or $n = m = p = q$ do not vanish. The last case represents a higher order of the Born approximation for the scattering of electrons and it may be omitted. Then c_4 is given by the first three cases:

$$c_4(\mathbf{k}', \mathbf{k}'', \mathbf{k}''', \mathbf{k}^{IV})\, \delta(\mathbf{k}' + \mathbf{k}'' + \mathbf{k}''' + \mathbf{k}^{IV}) =$$
$$= 2\pi^3 N(N-1) [\delta(\mathbf{k}' + \mathbf{k}'')\, \delta(\mathbf{k}'' + \mathbf{k}^{IV}) +$$
$$+ \delta(\mathbf{k}' + \mathbf{k}''')\, \delta(\mathbf{k}'' + \mathbf{k}^{IV}) + \delta(\mathbf{k}' + \mathbf{k}^{IV})\, \delta(\mathbf{k}'' + \mathbf{k}''')]. \quad (57.2)$$

If we substitute Eq. (57.2) into Eq. (56.16), the first of the three terms of Eq. (57.2) gives

$$G_0 \, (\mathbf{k}) \, \sum_2 \, (\mathbf{k}) \, G_0 \, (\mathbf{k}) \, \sum_2 \, (\mathbf{k}) \, G_0 \, (\mathbf{k}),$$

and, by definition given in Eq. (56.18), the rest represents Σ_4:

$$\sum_4 = \frac{N^2}{\Omega^2} \int | u \, (\mathbf{k}') |^2 \, | u \, (\mathbf{k}'') |^2 \, G_0 \, (\mathbf{k} - \mathbf{k}') \, G_0 \, (\mathbf{k} - \mathbf{k}' - \mathbf{k}'') \times$$
$$\times \, G_0 \, (\mathbf{k} - \mathbf{k}') \, d\mathbf{k}' d\mathbf{k}'' + \frac{N^2}{\Omega^2} \int | u \, (\mathbf{k}') |^2 \, | u \, (\mathbf{k}'') |^2 \times$$
$$\times \, G_0 \, (\mathbf{k} - \mathbf{k}') \, G_0 \, (\mathbf{k} - \mathbf{k}' - \mathbf{k}'') \, G_0 \, (\mathbf{k} - \mathbf{k}'') \, d\mathbf{k}' d\mathbf{k}''. \qquad (57.3)$$

These integrals, which are functions of $| \mathbf{k} |$ and E, have no poles and therefore Σ_4 gives only a small correction to the roots of the equation $1/\bar{G} = 0$.

However, if the term $G_0 \Sigma_2 G_0 \Sigma_2 G_0$ is not taken out of c_4, Σ_4 will have a term with $G_0 \, (\mathbf{k})$; this term becomes infinite when $E = \hbar^2 k^2 / 2m$ and very large when

$$E - \frac{\hbar^2 k^2}{2m} - \sum_2 \, (\mathbf{k}) = 0. \qquad (57.4)$$

In this case, the series (56.9) cannot be reduced to an approximate geometrical series (see below).

The quantity c_4 splits into three terms not only in the case of complete disorder, but also when there is a correlation between the atoms in a volume $\omega \ll \Omega$. If we fix $\mathbf{R}_n$, then the corrections to c_4 due to this correlation will be necessary if $| \mathbf{R}_n - \mathbf{R}_m |$ lies inside ω; these corrections are of the order of ω / Ω.

Thus, the main contribution to c_4 comes from the conditions when $\mathbf{k}' = -\mathbf{k}''$ and $\mathbf{k}''' = \mathbf{k}^{IV}$; the remaining terms are perturbations. Consequently, the possibility of replacing (56.9) with a geometric series depends not on the whole of c_4, but on the simplified expression

$$c_4' \, (\mathbf{k}', \ \mathbf{k}''') = \overline{\sum_{n, \, m, \, p, \, q} e^{i\mathbf{k}' \cdot (\mathbf{R}_n - \mathbf{R}_m) + \mathbf{k}''' \cdot (\mathbf{R}_p - \mathbf{R}_q)}}, \qquad (57.5)$$

$c_4 - c_4'$ is always small and will not be considered. The factorization condition is

$$c_4' \, (\mathbf{k}', \ \mathbf{k}''') = c_2 \, (| \mathbf{k}' | \, c_2 \, (| \mathbf{k}''' |). \qquad (57.6)$$

The integral (56.15) for Σ_2 depends on U and is complex. The sign of the imaginary part is the same as the sign of i^t, which determines the direction of by-passing a pole for $E = \hbar^2 k^2/2m$. We assume that

$$\Sigma_2 = A(\mathbf{k}, E) + i\Gamma(\mathbf{k}, E), \tag{57.7}$$

$$A = \frac{N}{(2\pi)^3 \Omega} \oint \frac{|u(\mathbf{k}')|^2}{E - \frac{\hbar^2}{2m}(\mathbf{k} - \mathbf{k}')^2} d\mathbf{k}', \tag{57.8}$$

$$\Gamma = \frac{N\pi}{(2\pi)^3 \Omega} \int |u(\mathbf{k}')|^2 \delta\left[E - \frac{\hbar^2}{2m}(\mathbf{k} - \mathbf{k}')^2\right] d\mathbf{k}'. \tag{57.9}$$

In the expression for A, we take the principal value of the integral. The electron mean free path depends only on Γ, and A gives the displacement of the level. As an example, we shall take

$$U(\mathbf{r}) = ua^3\delta(\mathbf{r}). \tag{57.10}$$

The transit time is $\tau = \hbar/\Gamma$ and the mean free path is $l = \tau\sqrt{\dfrac{2E}{m}}$. Substituting Eq. (57.10) into (57.9), we find that

$$\frac{1}{l} = \frac{1}{4\pi}\left(\frac{u}{E}\right)^2 \frac{a^6}{\lambda^4 r_0^3}, \tag{57.11}$$

where r_0 is the average distance between particles; λ is the electron wavelength $\lambda = \hbar/\sqrt{2mE}$. The smallness of Σ_4 is measured by the smallness of λ/l.

If the terms odd with respect to V are neglected, the Born approximation must be applied, i.e., the theory is not applicable in the general case.

Case of Complete Order

The value of c_2 for an ideal crystal is well known in connection with x ray diffraction:

$$c_2(\mathbf{k}) = \frac{N^2 (2\pi)^3}{\Omega} \sum_{\mathbf{K}} a(\mathbf{K})\delta(\mathbf{k} - \mathbf{K}), \tag{57.12}$$

where $\mathbf{K}$ represents reciprocal lattice vectors, $a(\mathbf{K})$ is the square of the modulus of the structure factor; for a Bravais lattice, $a = 1$ and we shall assume this here. The quaternary correlation function is given by the expression

$$\delta(\mathbf{k'} + \mathbf{k''} + \mathbf{k'''} + \mathbf{k^{IV}}) c_4(\mathbf{k'}, \mathbf{k''}, \mathbf{k'''}, \mathbf{k^{IV}}) = \frac{N^4(2\pi)^6}{\Omega^2} \times$$

$$\times \sum_{\mathbf{K'},\,\mathbf{K''},\,\mathbf{K'''}} \delta(\mathbf{k'} - \mathbf{K'})\,\delta(\mathbf{k''} - \mathbf{K''})\,\delta(\mathbf{k'''} - \mathbf{K'''})\,\delta(\mathbf{k'} + \mathbf{k''} + \mathbf{k'''} + \mathbf{k^{IV}}).$$

$$(57.13)$$

This sum includes the terms

$$\sum_{\mathbf{K'},\,\mathbf{K''}} \delta(\mathbf{k'} + \mathbf{k''})\,\delta(\mathbf{k'} - \mathbf{K'})\,\delta(\mathbf{k'''} + \mathbf{k^{IV}})\,\delta(\mathbf{k''} - \mathbf{K'''}),$$

which give the products of Σ_2. The remaining terms are not, in general, small and their influence can be estimated by considering the first approximation for $\overline{G}$. Since

$$\sum_2 = \frac{N^2}{\Omega^2} \sum_{\mathbf{k}} \frac{|u(\mathbf{K})|^2}{E - \dfrac{\hbar^2}{2m}(\mathbf{k} - \mathbf{K})^2}, \qquad (57.14)$$

in the first approximation

$$\overline{G} = \frac{1}{E - \dfrac{\hbar^2}{2m} k^2 + \dfrac{N^2}{\Omega^2} \displaystyle\sum_{\mathbf{k}} \dfrac{|u(\mathbf{K})|^2}{E - \dfrac{\hbar^2}{2m}(\mathbf{k} - \mathbf{K})^2}}. \qquad (57.15)$$

The above expression becomes infinite near the surface $E = \hbar^2 k^2/2m$ except where this surface intersects the surface $E = \dfrac{\hbar^2}{2m}(\mathbf{k} - \mathbf{K})^2$, giving rise to the usual forbidden band. Consequently, Eq. (56.18) contains all the forbidden bands which split the initial paraboloid $E = \hbar^2 k^2/2m$. However, the surfaces $(E, \mathbf{k})$ are periodic in $\mathbf{k}$ with a period equal to that of a reciprocal lattice; this periodicity is absent in Eq. (56.18) but it will appear if Σ_4 and higher-order terms are included.

Thus, the approximation represented by Eq. (57.6) is not completely satisfactory for crystals: it correctly predicts on the basis of Eq. (56.3) that the density of states is small everywhere apart from the unperturbed surface, but it cannot give further details about the density of states.

The situation is even more unsatisfactory in the intermediate cases which are the subject of our study.

§58. Electron Spectrum of Real Liquid Metals

In the preceding section, it was shown that in the case of a completely disordered substance $c_2(\mathbf{k}) = \text{const}$, but that for a crystal this quantity is a δ function series. Edwards thinks it is impossible to calculate theoretically c_2, c_4, etc. (obviously he was unaware of Soviet work on the liquid state theory). However, $c_2(\mathbf{k})$ can be found experimentally by x-ray or neutron diffraction. The experimental curve $c_2(\mathbf{k})$ for liquids differs greatly from that for a disordered substance case and has several sharp peaks which broaden on increase of $\mathbf{k}$.

The main difficulty in applying the theory to liquids is that the factorization $c'_4 = c_2 c_2$ is, in general, impossible. Thus we cannot carry out rigorous mathematical calculations and therefore Edwards has used additional, somewhat artificial assumptions.

It was assumed that $\overline{G}$ should be calculated in a region whose dimensions are of the order of the mean free path l. Then a considerable contribution to the sum (56.13) is made by terms with $|\mathbf{R}_n - \mathbf{R}_m| \leqslant l$, which can be represented mathematically by replacing (56.13) with the expression

$$c_2(\mathbf{k}) = \sum_{n,\,m} e^{i\mathbf{k}\cdot(\mathbf{R}_n - \mathbf{R}_m)} e^{-\frac{\mathbf{R}_n - \mathbf{R}_m}{l}}. \tag{58.1}$$

Physically it means that an electron is described by a plane wave which decays at a distance l.

A quantity R is now introduced, which denotes the distance over which order is retained in the distribution of atoms. In Chap. IV, we showed that $R \simeq r_1/\varepsilon^2$, where ε is the degree of departure from short-range order.

All liquids may be divided into four classes in terms of the parameters l and R.

1. $R \simeq r_1$. This is the case of complete disorder applicable to gases or to liquid metals near the critical temperature. Here we may partly use the calculation in § 57; in particular, $\overline{G}$ is found from

$$\left[E - \frac{\hbar^2 k^2}{2m} + \sum_2(k)\right]\overline{G} = 1. \tag{58.2}$$

However, in contrast to Eq. (57.1), we must use an experimentally determined value of $c_2(k)$. Obviously, c_2 is a function only of the modulus of $\mathbf{k}$.

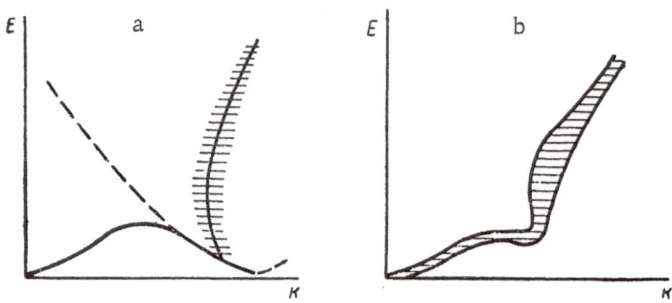

Fig. 24. Projection of $\rho(E, k)$ on to the (E, k) plane for a disordered liquid metal: a) $c_2(k) = \delta(K)$ case; b) $c_2(k)$ maximum
spread over a range of k.

2. $l \gg R \gg r_1$. This condition is associated with the use of the nearly
free electron method; when the perturbation potential is weak the scattering is
also weak, irrespective of the degree of order, and l is large. Between two
collisions, an electron moves a distance over which long-range order is absent.
In Eq. (58.1), we can drop the real exponential term and the present case reduces to class 1.

3. $R \gg l \simeq r_0$. When this condition applies, the electron scattering is
strong so that the nearly free electron approximation is inapplicable. Only
the short-range order is important and this order is always retained.

4. $l \simeq R$ or $R \gg l \gg r_1$. This class is typical of real liquids, in particular, many liquid metals. Here $c'_4 \neq c_2 \cdot c_2$ and a special treatment is needed,
which unfortunately cannot be carried out rigorously. The limit of this case
is the "polycrystalline model," according to which a liquid consists of many
ordered regions which are oriented at random with respect to one another. This
model is similar to the "crystallite" glass structure discussed in § 21.

Thus, classes "1" and "2" can be unified and called the case of strongly
disordered liquids, while class "4" is the ordered liquid case. There is no point
in dealing with class "3" by the weakly interacting electron method.

In strongly disordered liquid metals, $\overline{G}$ is given by Eq. (58.2). Then, according to Eq. (56.15),

$$\Sigma_2 (\mathbf{k}) = \frac{1}{(2\pi)^3 \Omega} \int \frac{| u (\mathbf{k}') |^2 c_2 (\mathbf{k}')}{E - \frac{\hbar^2}{2m} (\mathbf{k} - \mathbf{k}')^2 + i\iota} d\mathbf{k}' =$$

$$= \frac{1}{4\pi^2\Omega} \int\limits_0^\infty |u(\mathbf{k'})|^2 c_2(\mathbf{k'}) \int\limits_{-1}^1 \frac{k'^2 dk d\mu}{E - \frac{\hbar^2}{2m}(k^2 + k'^2 - 2kk'\mu) + i\iota} =$$

$$= \frac{m}{\hbar^2 kk'} \cdot \frac{1}{4\pi^2\Omega} \int\limits_0^\infty |u(k')|^2 c_2(k') \ln \frac{E - \frac{\hbar^2}{2m}(k-k')^2}{E - \frac{\hbar^2}{2m}(k+k')^2} k'^2 dk' +$$

$$+ \frac{i}{4\pi\Omega} \int\limits_0^\infty |u(k')|^2 c_2(k') \times$$

$$\times \theta \left\{ \left[E - \frac{\hbar^2}{2m}(k^2 + k'^2) \right] \frac{m}{\hbar^2 kk'} \right\} \frac{m}{\hbar^2 kk'} k'^2 dk',$$

where

$$\theta(x) = 1 \quad (-1 < x < 1),$$
$$= 0 \quad (1 < x, \; x < -1), \tag{58.3}$$

and $c_2(k)$ should be found by experiment. As a rough approximation, we shall assume that $c_2(k) = \delta(K)$. Combining $|u(K)|^2$, the density, and the numerical coefficient into a single constant ζ, we obtain

$$\Sigma_2(k) = \frac{\zeta}{2Kk} \ln \left| \frac{E - \frac{\hbar^2}{2m}(k+K)^2}{E - \frac{\hbar^2}{2m}(k-K)^2} \right| +$$

$$+ \frac{i\pi\zeta}{2Kk} \theta \left\{ \left[E - \frac{\hbar^2}{2m}(k^2 + K^2) \right] \frac{m}{\hbar^2 Kk} \right\}. \tag{58.4}$$

Having obtained an expression for Σ_2, we can easily calculate $\rho(E, k)$ using Eqs. (58.2) and (56.3). Figure 24a shows the projection $\rho(E, k)$ onto the (E, k) plane.

Below the $E = \frac{\hbar^2}{2m}(K-k)^2$ parabola, shown dashed, the imaginary part of Σ_2 is zero, so that

$$\rho(E, k) = \delta \left[E - \frac{\hbar^2}{2m} k^2 + \Sigma_2(k) \right], \tag{58.5}$$

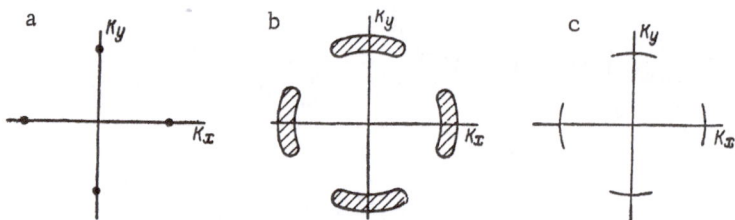

Fig. 25. The correlation function $c_2(\mathbf{k})$ for two dimensions: a) a crystal; b) an ordered liquid; c) a simplified model of an ordered liquid.

and the continuous curve in Fig. 24a is given by

$$E = \frac{\hbar^2}{2m} k^2 + \frac{\zeta}{2Kk} \ln \left| \frac{E - \frac{\hbar^2}{2m} (K - k)^2}{E - \frac{\hbar^2}{2m} (K + k)^2} \right|. \qquad (58.6)$$

Above the $E = \frac{\hbar^2}{2m}(K - k)^2$ parabola, where the imaginary part of Σ_2 is not equal to zero, we have a diffuse band instead of a curve, the center of this band being close to the initial parabola $E = \frac{\hbar^2}{2m} k^2$. With increase of k, the width of this band tends to zero so that it reduces to the initial parabola. The energy spectrum is very different from the crystalline case; in particular, there is no forbidden band. This is solely due to the lack of a dependence of c_2 on the direction of $\mathbf{k}$.

We shall assume now that $c_2(k) \neq \delta(K)$, but that it is represented by a more or less diffuse peak around K. In this case, the lower curve also spreads into a band and the whole band straightens somewhat, assuming the form shown in Fig. 24b. As the peak broadens, the average contour of the band approaches more and more closely the initial parabola $E = \frac{\hbar^2}{2m} k^2$.

We shall now consider the case of ordered liquid metals. This case is most difficult to treat mathematically but it is of the greatest interest since it allows us to determine how the E, $\mathbf{k}$ curve of a crystal changes into the curves of Fig. 24 when the degree of order is reduced.

We shall calculate c_2 allowing for local crystalline regions extending over a distance l or more. Summing Eq. (58.1) for m only, we obtain the function

$$g_2(\mathbf{R}_n,\ \mathbf{k}) = \sum_m e^{i\mathbf{k}\ \cdot\ (\mathbf{R}_n-\mathbf{R}_m)-\frac{|\mathbf{R}_n-\mathbf{R}_m'|}{l}}, \tag{58.7}$$

which has broad peaks in the **k**- space which are oriented in definite directions.

Then

$$c_2(\mathbf{k}) = \sum_n g_2(\mathbf{R}_n,\ \mathbf{k}). \tag{58.8}$$

It is assumed that the short-range order is everywhere the same and that there is a system of crystallographic axes, which rotates with some correlation on moving from atom to atom. This correlation is given by the correlation function c_4'

$$c_4'(\mathbf{k},\ \mathbf{k}') = \overline{\sum_{n,\ m} g(\mathbf{R}_n,\ \mathbf{k})\,g(\mathbf{R}_m,\ \mathbf{k}')}. \tag{58.9}$$

Figure 25a shows $c_2(\mathbf{k})$ for a crystal: it has a nonzero value at four points shown in this figure. A solution for a polycrystalline substance is obtained by carrying out a calculation using this $c_2(\mathbf{k})$ and then averaging the final result over all orientations of the axes. However, for a liquid — even an ordered one — we must assume that $c_2(\mathbf{k})$ has broad peaks, as shown in Fig. 25b, where $c_2 \neq 0$ in the hatched regions, (this represents a binary distribution function of the (18.11) type). This value of c_2 should be substituted into the integral of Eq. (56.14), which we shall denote by Σ_2'. The peaks in Fig. 25b

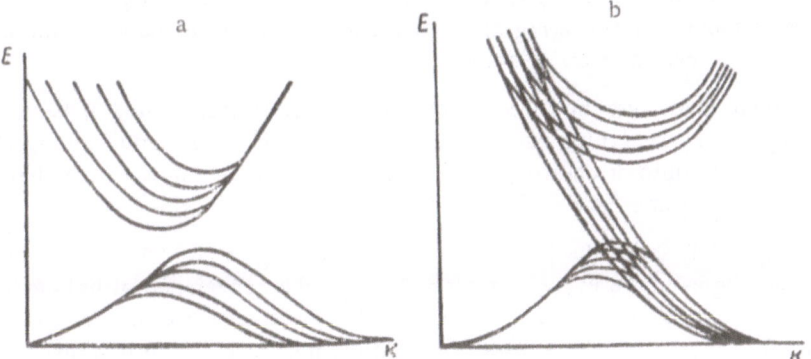

Fig. 26. E(k) curves: a) for a polycrystalline substance; b) for a simplified model of an ordered liquid.

will be denoted by $\mathbf{\dot{K}}_j$, so that $g_2 = g_2(\mathbf{k}, \mathbf{K}_j)$. In the calculation of $\overline{G}$ we must average out over all possible orientations of $\mathbf{K}_j$. Consequently, we obtain, as in Eq. (56.19),

$$\overline{G} = \int (G_0 + G\Sigma_2'G_0 + G_0\Sigma_2'G_0\Sigma_2'G_0 + \ldots)d\omega_j, \qquad (58.10)$$

where $d\omega_j$ is an element of the solid angle in the $\mathbf{K}_j$ space, and

$$\Sigma_2' = \int G_0(\mathbf{k} - \mathbf{k}')\, g_2(\mathbf{k}', \mathbf{K}_j)\, d\mathbf{k}'. \qquad (58.11)$$

Then

$$\int \Sigma_2' d\omega_j = \int G_0(\mathbf{k} - \mathbf{k}')\, g_2(\mathbf{k}', \mathbf{K}_j)\, d\mathbf{k}' =$$

$$= \int G_0(\mathbf{k} - \mathbf{k}')\, c_2(\mathbf{k}')\, d\mathbf{k}' \qquad (58.12)$$

and

$$\int \Sigma_2' G_0 \Sigma_2' d\omega_j = \int G_0(\mathbf{k} - \mathbf{k}')\, g_2(\mathbf{k}', \mathbf{K}_j)\, G_0(\mathbf{k} - \mathbf{k}'') \times$$

$$\times\, g_2(\mathbf{k}', \mathbf{K}_j)\, d\mathbf{k}'d\mathbf{k}''d\omega_j = \int G_0(\mathbf{k} - \mathbf{k}')\, G_0(\mathbf{k} - \mathbf{k}'') \times$$

$$\times\, c_4'(\mathbf{k}', \mathbf{k}'')\, d\mathbf{k}'d\mathbf{k}''. \qquad (58.13)$$

The equations (58.12) and (58.13) follow from the fact that integration with respect to angle in the $\mathbf{K}_j$ space is equivalent to averaging over $\mathbf{R}_n$ in Eqs. (58.8) and (58.9).

Using these formulas, calculations have been carried out for a somewhat simplified model, shown in Fig. 25c. Here, the peaks are δ shaped in the k but are spread over a range of angles.

Figure 26a shows the (E, k) curves for the polycrystalline model, deduced from Fig. 25a by averaging over the orientations of $\mathbf{K}$. Each curve in Fig. 26a represents a definite orientation of $\mathbf{K}$. These curves fill completely the band between the extreme curves of each set.

Figure 26b shows the (E, k) curves corresponding to Fig. 25c. The scatter of the angles in the expression for c_2 leads to the appearance of linking curves between the corresponding curves of the upper and lower bands; the lower band curves come closer together and the upper band is shifted upward toward higher energies. When the angular scatter increases, the pattern approaches that shown in Fig. 24a, where the upper band has already disappeared at infinity.

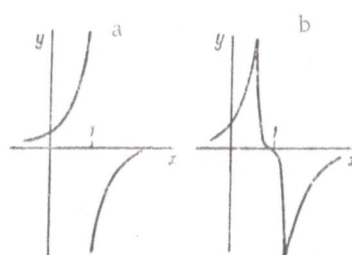

Fig. 27. Function $y(x)$: a) $y = 1/(1-x)$;
b) y averaged out over a certain range.
A linking curve appears in the latter
case, joining the two branches of the
curve a.

The appearance of the linking curves can be understood using the following simple example [1]. We shall consider the function

$$y = \frac{1}{1-x}.$$

It is shown in Fig. 27a and has two branches. We shall now assume that
y is averaged out in accordance with

$$\bar{y} = \frac{a}{2\pi} \int\limits_{-\alpha}^{\alpha} \frac{d\varepsilon}{1-x+\varepsilon} = \frac{1}{2\alpha} \ln \left| \frac{1-x-a}{1-x+a} \right|.$$

The plot of $\bar{y}$, shown in Fig. 27b, has a curve which joins the two branches.
Similarly, if instead of a discrete value of K_j, representing a crystal, we
substitute in Eq. (58.11) a value spread over a certain range of angles (Fig.
25c), we shall obtain the linking curves in the (E, k) plot.

The presence of the linking curves destroys the absolutely forbidden band
but, since the density of states along these curves is lower than along the main
curves, a band with a lower density of levels is formed. This is in complete
agreement with the conclusions presented in § 28 for the case of narrow
forbidden bands using the results of the calculations carried out in Chap. IV.

We must stress once again that this pattern of the spectrum applies to
weakly bound electrons. If the electron binding to the atoms is stronger, the
linking curves of Fig. 26b may close without forming a bridge between the

bands [1]. In this case, an absolutely forbidden band is possible in a liquid conductor, in agreement with the results given in Chaps. IV and V for broad forbidden bands, and with the results of the numerical calculations presented in § 28.

§59. Wave Functions of Electrons in Liquid Metals [7]

In the present chapter, we have not yet considered the problem of the electron functions because we have calculated the density of states directly, using the Green's functions. A study of the wave functions for liquids using the model of weakly bound electrons has been carried out by Heine [7].

Heine assumed plane waves and used a pseudopotential established by the atoms as a perturbation. The potential of real atoms is far too strong to be considered as a perturbation, but a considerable part of the atomic potential is compensated by the electron kinetic energy by virtue of the Pauli principle. The resultant pseudopotential is relatively weak and can justifiably be regarded as a perturbation

$$V(\mathbf{r}) = \sum_n U(\mathbf{r} - \mathbf{R}_n);\qquad(59.1)$$

the notation is the same as in Eq. (56.1).

The electron energy in the second approximation of the perturbation theory is (in atomic units: $\hbar = m = e = 1$):

$$E(\mathbf{k}) = \frac{k^2}{2} + V_{\mathbf{kk}} + 2 \sum{}' \frac{V_{\mathbf{k'k}}}{k^2 - k'^2}.\qquad(59.2)$$

The corresponding wave function is

$$\psi = \frac{1}{\sqrt{\Omega}} e^{i\mathbf{k}\cdot\mathbf{r}} + \frac{2}{\sqrt{\Omega}} \sum{}' \frac{V_{\mathbf{k'k}}}{k^2 - k'^2} e^{i\mathbf{k}\cdot\mathbf{r}},\qquad(59.3)$$

where Ω is the volume of the whole system, Σ' denotes summation for $\mathbf{k'} \neq \mathbf{k}$, $V_{\mathbf{k'k}}$ is a matrix element of the pseudopotential.

The function (59.3) is unsatisfactory for two reasons. First, it is not a Bloch wave propagated in a liquid and it contains the scattered wave. Secondly, if we attempt to normalize Eq. (59.3), we obtain a diverging sum

$$\sum{}' \frac{|V_{\mathbf{k'k}}|^2}{(k^2 - k'^2)^2}.\qquad(59.4)$$

Therefore, it is necessary to subtract from Eq. (59.3) the scattered wave and obtain a quasi-Bloch function, similar to Eq. (35.2), used in Chaps. VI and VII. The scattered and quasi-Bloch functions cannot be separated explicitly in the wave function itself but this can be done in any matrix element used in calculating the various electronic properties of liquid metals; the quasi-Bloch functions are not eigenfunctions of the stationary Schrödinger operator but they have a sufficiently long lifetime to be used in the theory of time-dependent perturbations.

Using this theory, we shall state a perturbation $V(\mathbf{r})\,e^{at}$ at a moment $t = -\infty$. At $t = 0$, the wave function is

$$\psi = \frac{1}{\sqrt{\Omega}}\,e^{i\mathbf{k}\cdot\mathbf{r}} + \frac{2}{\sqrt{\Omega}}\sum{}'\,\frac{V_{\mathbf{k'k}}}{k^2 + k'^2 + 2ia}\,e^{i\mathbf{k}\cdot\mathbf{r}}. \qquad (59.5)$$

The rate of application of the perturbation a is assumed to be small (adiabatic condition).

The function (59.5) includes both the scattered wave and the quasi-Bloch wave, which we shall denote by φ.

We shall show how to subtract the scattered wave in calculating any matrix element, using as an example the calculation of the normalizing integral $\int |\varphi|^2\,d\mathbf{r}$. The total normalizing integral for the wave function ψ is

$$\int |\psi|^2\,d\mathbf{r} = 1 + \sum{}'\,\frac{|V_{\mathbf{k'k}}|^2}{\frac{1}{4}(k^2 + k'^2)^2 + a^2}. \qquad (59.6)$$

Summation is replaced by integration with respect to k and $d\omega_{k'}$. We shall consider separately the contribution of the region near $k' = k$, writing

$$\int |\psi|^2\,d\mathbf{r} = 1 + \frac{\Omega}{(2\pi)^3}\int \frac{|V_{\mathbf{k_1 k}}|^2\,k_1 k'\,dk'\,d\omega_{k'}}{\frac{1}{4}(k^2 - k'^2)^2 + a^2} +$$

$$+ \frac{\Omega}{(2\pi)^3}\int \frac{[|V_{\mathbf{k'k}}|^2\,k' - |V_{\mathbf{k_1 k}}|^2\,k_1]\,k'\,dk'\,d\omega_{k'}}{\frac{1}{4}(k^2 - k'^2)^2 + a^2}; \qquad (59.7)$$

here k_1 is a wave vector with the same direction as k' but of magnitude $k_1 = k$, i.e., it is a singularity in the integration with respect to k'. In the second term of Eq. (59.7), the integration with respect to k' can be carried out rigorously if we introduce the variable $x = 1/2\,(k'^2 - k^2)$. We obtain

$$\frac{\Omega}{(2\pi)^3} \int d\omega_{k'} |V_{k,k}|^2 k_1 \int\limits_{-\frac{1}{2}k^2}^{\infty} \frac{dx}{x^2 + a^2} \; ;$$

here the first integral is equal to $(\pi\tau)/2$, where τ is the relaxation time due to scattering. The second integral is

$$\frac{1}{a} \left(\pi - \tan^{-1} \frac{2a}{k^2} \right) = \frac{\pi}{a} - \frac{2}{k^2} + 0(a^2).$$

In the last integral of Eq. (59.7), we expand the denominator in series of powers of a. Consequently, Eq. (59.7) becomes:

$$\int |\psi|^2 d\mathbf{r} = \frac{1}{2a\tau} + \left\{ 1 - \frac{2}{k^2} \cdot \frac{\Omega}{(2\pi)^3} \int |V_{k,k}|^2 k_1 d\omega_{k'} + \right.$$
$$\left. + \frac{\Omega}{(2\pi)^3} \int \frac{[|V_{k'k}|^2 k' - |V_{k,k}|^2 k_1] k' dk' d\omega_{k'}}{\frac{1}{4}(k^2 - k'^2)^2} \right\} + 0\left(\frac{V^2 a^2}{k^8} \right). \quad (59.8)$$

The first term represents the scattered wave, since for an exponential dependence of the perturbation on time—of the exp(at) type—the effective duration of the perturbation is 1/a and the scattered part of the wave should be of the order of 1/a. The last term is due to the process of applying the perturbation and it tends to zero when the rate of application tends to zero. Thus the normalizing integral of the quasi-Bloch wave is given by the middle part of Eq. (59.8):

$$\int |\varphi|^2 d\mathbf{r} = 1 - \frac{2}{k^2} \frac{\Omega}{(2\pi)^3} \int |V_{k,k}|^2 k_1 d\omega_{k'} +$$
$$+ \frac{\Omega}{(2\pi)^3} \int \frac{[|V_{k'k}|^2 k' - |V_{k,k}|^2 k_1] k' dk' d\omega_{k'}}{\frac{1}{4}(k^2 - k'^2)} . \quad (59.9)$$

Similarly, a matrix element of any operator $\hat{P}$, containing functions φ_k, may be calculated by writing down a matrix element containing functions ψ in the form of:

$$\int \psi^* \hat{P} \psi d\mathbf{r} = \frac{A}{a} + B + 0(a^2) \qquad (59.10)$$

and taking the middle part. Having carried out these calculations, we can show that, for example, the energy E(k) found from the formula

$$E(\mathbf{k}) = \frac{\int \varphi_k^* \hat{H} \varphi_k d\mathbf{r}}{\int \varphi_k^2 d\mathbf{r}}, \qquad (59.11)$$

where $\hat{H} = \frac{1}{2} \nabla^2 + V(\mathbf{r})$, is identical with Eq. (59.2) up to second-order perturbations. Similarly, we can show that the electron velocity, defined as

$$v = \frac{-i \int \varphi_k \nabla \varphi_k d\mathbf{r}}{\int |\varphi_k|^2 d\mathbf{r}}, \qquad (59.12)$$

is equal to $\partial E / \partial k$.

Heine [7] applied his method to the calcuation of the resistance and the Knight shift in liquid metals (the Knight shift is the change in the n.m.r. frequency in a metal compared with a nonmetallic compound containing this metal [8]).

Chapter XI

THEORY OF THE ELECTRICAL CONDUCTIVITY
OF LIQUID METALS [1, 2]

In Chap. VIII, we calculated the various galvanomagnetic coefficients of liquid and amorphous substances, considering liquids as distorted crystals. We can also deal with liquid metals by considering them as gases. Although the experimental observations presented in Chap. III show that liquids are closer in their structure to crystals than to gases, an approach to the problem of liquid structure is from another viewpoint of great interest.

The electrical properties of liquid metals were calculated by Ziman[1] in the weakly bound electron approximation using the quasi-gas model. In contrast to the theory presented in Chap. X, Ziman did not consider the energy spectrum and electron eigenfunctions, but deduced all the electrical properties of liquid metals from the scattering of plane waves by metal atoms. Obviously, this approach is possible for metals but not semiconductors. However, the advantage of Ziman's treatment is that all the parameters occurring in it can be found by independent experiments.

The scattering of the plane waves $\psi_{\mathbf{k}} = \frac{1}{\sqrt{\Omega}} e^{i\mathbf{k}\cdot\mathbf{r}}$ is considered using the atomic pseudopotential (§ 59) as the scattering potential. The total scattering potential is

$$V(\mathbf{r}) = \sum_n U(\mathbf{r} - \mathbf{R}_n).$$

(60.1)

In applications, it is necessary to know the matrix element of this potential between two unperturbed states

$$V_{\mathbf{k}''\mathbf{k}'} = \frac{1}{\Omega} \int V(r) e^{i(\mathbf{k}'-\mathbf{k}'')\cdot\mathbf{r}} d\mathbf{r} = U_{\mathbf{k}'-\mathbf{k}''} \sum_n e^{i(\mathbf{k}'-\mathbf{k}'')\cdot\mathbf{R}_n},$$

(60.2)

234

where the Fourier transform of the potential of a single atom is

$$U_k = \frac{N}{\Omega} \int e^{i\mathbf{k}\cdot\mathbf{r}} U(\mathbf{r}) \, d\mathbf{r}. \tag{60.3}$$

It is difficult to calculate this quantity theoretically, but for one value of $\mathbf{k}$, i.e., for $\mathbf{k} = \mathbf{K}$, where $\mathbf{K}$ is one of the vectors of the reciprocal lattice of the metal in the corresponding crystal state, $U_\mathbf{K}$ can be estimated from the experimental data.

According to the theory of weakly bound electrons, the forbidden band width is

$$E_g = 2|U_\mathbf{K}|. \tag{60.4}$$

In univalent metals, the forbidden band is formed near the plane to the center of a Brillouin zone [(111) or (110), depending on the crystal structure], which determines the form of the Fermi surface.

The forbidden band width may be found also from transport phenomena [3-5].

Thus for each univalent metal we may assume that we know approximately one Fourier component of the pseudopotential. In order to estimate U_k for any value of $\mathbf{k}$, we shall assume that the dependence on $\mathbf{k}$ is approximately the same for all metals. According to the calculations of Phillips and Kleinman [6], the components U_{111} and U_{200} are quite large for silicon but they decrease at high values of $\mathbf{k}$.

Ziman assumes initially that in a liquid metal there is no order of any kind (the gas model) and ions scatter electrons independently of one another. Then the electrical conductivity, as in the ionized gases, is

$$\sigma = \frac{ne^2 l}{m v_F}, \tag{60.5}$$

where n is the electron density, v_F is the velocity near the Fermi level, and l is given by

$$\frac{1}{l} = \frac{N_i}{\Omega} 2\pi \int_0^{2\pi} (1 - \cos\vartheta) f(\vartheta) \sin\vartheta \, d\vartheta, \tag{60.6}$$

where N_i is the concentration of ions, and $f(\vartheta)$ is the differential cross section for ion scattering.

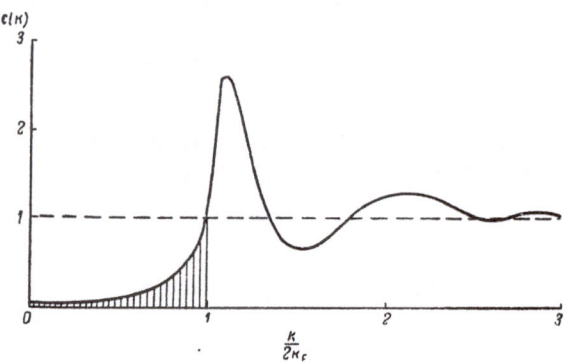

Fig. 28. Form of the correlation function c(k) for liquid rubidium from neutron-diffraction data.

It is assumed that the cross section may be calculated in the Born approximation, being due to the ion pseudopotential $U(r-R_n)$. The quantum
theory of collisions gives

$$f(\vartheta) = \frac{2\pi}{\hbar v_F} \left| \frac{1}{N} U_k \right|^2 \frac{3}{4} \cdot \frac{n}{E_F} \cdot \frac{1}{4\pi} .$$ (60.7)

Here $\mathbf{k} = \mathbf{k'} - \mathbf{k''}$ is the scattering vector whose value varies from
0 to $2k_F$.

Substituting Eq. (60.7) into Eq. (60.6), we obtain

$$l_{gas} = \frac{2\sqrt{2}}{3\pi\sqrt{E_F}} \left| \frac{E_F}{U} \right|^2 ,$$ (60.8)

$$|U|^2 = \frac{1}{4k_F^4} \int_0^{2k_F} |U_k|^2 k^3 dk.$$ (60.9)

In order to calculate these expressions, it is necessary to know U_k as a
function of k, but we know only U_K, where K is slightly greater than $2k_F$. We
shall assume that U_k is on the average equal to U_K and substitute the latter
in Eq. (60.9) in place of U_k.

The values of l_{gas} calculated in this way [1] for all metals are, except
for sodium, much smaller than the experimentally determined electron mean
free path in liquid metals [7]. The reason for this discrepancy lies in the

incorrect assumption of the absence of order in liquid metals and in the disregard of the correlation between the scattering of electrons by various ions. This correlation obviously reduces the scattering since a completely ordered ideal crystal does not scatter at all any electron waves at T = 0.

To allow for the correlation of scattering in a liquid, it is necessary to calculate the scattering not by individual ions but by the potential (60.2). The sum in Eq. (60.2) is irregular and has a random phase. However, we need only the average value of the square of the modulus of the matrix element of the potential (60.2), which can be expressed in terms of the density of atomic distribution in a liquid $\rho(r)$ (see § 18). Unit volume is taken to be that volume which is occupied on the average by one atom, so that $\overline{\rho(r)} = 1$ and $\rho(r) \to 1$ as $r \to \infty$. In considering x-ray diffraction [8], it has been shown that

$$c(k) \equiv \left| \frac{1}{N} \sum_n e^{i\mathbf{k}\cdot\mathbf{R}_n} \right|^2 = 1 + \frac{4\pi N}{\Omega} \int\limits_0^\infty [\rho(r) - 1] \frac{\sin kr}{kr} r^2 dr. \tag{60.10}$$

Allowance for the correlation in a liquid reduces to the substitution of the potential (60.2) in Eq. (60.7) in place of U_k, i.e., it is necessary to divide l_{gas} by a quantity $\bar{c}$ which represents $c(k)$ averaged out over k using the relationship

$$\bar{c}|U|^2 = \frac{1}{4k_F^4} \int\limits_0^{2k_F} |U_k|^2 c(k) k^3 dk; \tag{60.11}$$

then

$$l_l = \frac{l_{gas}}{\bar{c}} = \frac{2\sqrt{2}E_F^{3/2}}{3\pi\bar{c}|U|^2}. \tag{60.12}$$

In a gas $c(k) = 1$, but in a liquid $c(k)$, according to x-ray and neutron diffraction studies [9], has the form shown in Fig. 28. At wavelengths approximately equal to the average interatomic distance, there is a sharp maximum. At lower values of k, the quantity $c(k)$ decreases rapidly. Since the maximum lies at $k_m > 2k_F$, $c(k) < 1$ in the integration interval of Eq. (60.1), shown hatched in Fig. 28. Analyzing this curve and using some theoretical reasoning (which does not seem fully convincing), Ziman made the reasonable assumption that

$$\bar{c}|U|^2 = \alpha|U_K|^2, \tag{60.13}$$

where α lies between 0.05 and 0.2. In other words, the correlation increases the mean free path of electrons in liquid metals by a factor of 5-20. This makes it possible to obtain agreement between the values of l calculated using Eq. (60.12) and the experimental values for liquid univalent metals, except for sodium. However, we cannot say that the agreement is quantitative until we know the exact values of α for real metals. Moreover, the latest x-ray and neutron diffraction studies give $\alpha = 0.4$ [1], which makes the agreement between theory and experiment poorer.

§61.   Plasma Scattering and the Displacement Potential

The theory developed in the preceding section does not give the correct electron mean free path for liquid sodium because for this metal $l_{obs}/l_{gas} = 0.5$.

To explain this, it is suggested that liquids contain many vacant sites which scatter conduction electrons. A vacant site represents the absence of a positive ionic charge, and therefore it repels and scatters electrons. This scattering is not included in the scattering by the pseudopotential and it may be important, particularly in the case of sodium, for which $U_{\mathbf{K}}$ is small.

To calculate this effect, it is necessary to consider local density fluctuations in a liquid in the same way as was done in § 53. We shall denote a local dilatation by $\Delta(\mathbf{r})$. To determine the scattering, we need the average quadratic Fourier component

$$|\Delta_{\mathbf{k}}|^2 = c(\mathbf{k}),\qquad(61.1)$$

where $c(\mathbf{k})$ is the function which we know from Eq. (60.10).

In a metal, in contrast to a semiconductor, a dilatation does not represent a scattering region with a charge $ne\Delta$. The system of ions and electrons forms a plasma in which the density fluctuations of the heavy ions are screened by electrons. However, in a degenerate Fermi plasma this screening is incomplete because it gives rise to oscillations of the Fermi level. The effective potential associated with a dilatation is the plasma potential

$$W_0\Delta = \frac{2}{3}\,E_{\mathrm{F}}\Delta.\qquad(61.2)$$

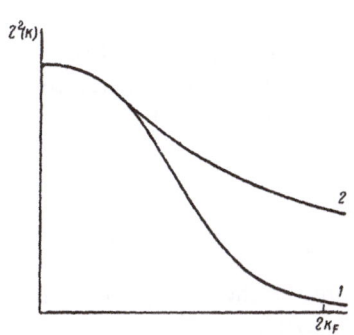

Fig. 29. Square of the matrix element of the scattering potential Z as a function of k: 1) plasma potential; 2) pseudopotential.

In the limiting case of long wavelengths, the principle of equal energy distribution over classical degrees of freedom gives

$$c(0) = \frac{k_B TN}{\Omega} \beta_l, \qquad (61.3)$$

where k_B is Boltzmann's constant, and β is the compressibility.

It follows from Eqs. (61.1)-(61.3) that the probability of a transition in the plasma scattering at long wavelengths is proportional to

$$|W_0|^2 c(0) = \left(\frac{2}{3} E_F\right)^2 \frac{k_B TN\beta}{\Omega}.$$

$$(61.4)$$

Unfortunately, W_k cannot be calculated exactly for large values of k, but obviously it should decrease rapidly with increase of k. An approximate calculation has been carried out for a model in which the electron charge is uniformly distributed over a sphere of radius r_1 which encloses one atom.

In this case

$$W_k \simeq \frac{2}{3} E_F G(kr_1), \qquad (61.5)$$

where

$$G(x) = \frac{3x \cos x - \sin x}{x^3}. \qquad (61.6)$$

Averaging out W_k with a weighting coefficient $k^3 c(k)$, as in Eq. (60.11), we obtain

$$l_{\text{plasma}} = \frac{2\sqrt{2}}{3\pi \sqrt{E_F}} \left(\frac{E_F}{\frac{2}{3} E_F}\right)^2 \frac{1}{\overline{c G^2}}. \qquad (61.7)$$

A rough estimate gives $\overline{G^2} \simeq 0.1$. For sodium, Eq. (61.7) gives the correct order of magnitude of the mean free path. Thus, in the case of sodium, the plasma scattering is dominant. However, this scattering does occur in other univalent metals along with the scattering on the pseudopotential. We may assume that the value of $\overline{c G^2}$ is the same for all liquid univalent metals. Then, according to Eq. (61.7), $l_{plasma}\sqrt{E_F}$ = const; using the experimental value of this constant for sodium, we can calculate the plasma mean free path for other liquid metals. Subtracting the reciprocal of the plasma mean free path from the reciprocal of the experimentally determined path, Ziman found the remainder $1/l_{str}$, where l_{str} is the mean free path due to the structural scattering:

$$\frac{1}{l_{str}} = \frac{1}{l_{obs}} - \frac{1}{l_{plasma}}. \qquad (61.8)$$

The values of l_{str} calculated in this way are, for the majority of metals, approximately 20 times larger than l_{gas}, i.e., $\bar{c} \approx 1/20$.

We can combine Eqs. (60.12) and (61.7) by writing

$$l = \frac{2\sqrt{2E_F^{3/2}}}{3\pi\bar{c}\,\overline{Z^2}}. \qquad (61.9)$$

Here we have introduced a new function

$$|Z(k)|^2 \equiv |U_k|^2 + |W_k|^2 \qquad (61.10)$$

the average value of which is given by the expression

$$\overline{Z^2}\bar{c} = \frac{1}{4k_F^4} \int_0^{2k_F} |Z(k)|^2 c(k) k^3 dk; \qquad (61.11)$$

$Z(k)$ combines the pseudopotential and the plasma potential, and since both these potentials are due to the displacement of ions, we can call it the displacement potential.

Figure 29 shows the expected behavior of $|Z(k)|^2$ for the plasma potential and the pseudopotential. In accordance with our treatment, $|Z(k)|^2$ for the plasma potential decreases more rapidly with k than does the same quantity for the pseudopotential.

§ 62.   Change in the Electrical Conductivity on Melting

The problem of the change in the conductivity of metals on melting has been discussed in §1. Table 1 lists the experimental values of σ_s/σ_l for several metals, as well as the values of σ_s/σ_l calculated on the assumption that only the phonon scattering of the electrons changes on melting. The theory given in the present chapter yields a more general explanation of the change in the electrical conductivity on melting. To arrive at this explanation, we shall carry out calculations for a solid in the same way as for a liquid, assuming for both cases the same function $Z(k)$ which represents the effective potential for the scattering of the electrons.

The only item which changes in Eqs. (61.9) and (61.11) is the correlation function $c(k)$. For a solid, this function may be calculated directly from the definition in Eq. (60.10):

$$c_S(\mathbf{k}) = \left| \frac{1}{N} \sum_n e^{i\mathbf{k} \cdot \mathbf{R}_n} \right|^2. \tag{62.1}$$

Assume now that in an ideal lattice an ion has a radius vector $\mathbf{R}_n^0$ and that under the influence of thermal vibrations of the lattice the ion is displaced so that its radius vector is

$$\mathbf{R}_n = \mathbf{R}_n^0 + \sum_q \mathbf{y}_q e^{-i\mathbf{q} \cdot \mathbf{R}_n^0}, \tag{62.2}$$

where $\mathbf{q}$ and $\mathbf{y}_q$ are, respectively, the wave vectors and amplitudes of the thermal waves. Assuming that $\mathbf{y}_q$ is small, we expand Eq. (62.1) in powers of $\mathbf{y}_q$:

$$c_S(\mathbf{k}) = \left| \frac{1}{N} \sum_n e^{i\mathbf{k} \cdot \mathbf{R}_n^0} \left[1 + i\mathbf{k} \sum_q \mathbf{y}_q e^{-i\mathbf{q} \cdot \mathbf{R}_n^0} \right] \right|^2 =$$

$$= \left| \delta_{\mathbf{k}, \mathbf{K}} + i \sum_q (\mathbf{y}_q \cdot \mathbf{k}) \delta_{\mathbf{k}-\mathbf{q}, \mathbf{K}} \right|^2 = \delta_{\mathbf{k}, \mathbf{K}} + \left| \mathbf{k} \mathbf{y}_{\mathbf{k}-\mathbf{K}} \right|^2; \tag{62.3}$$

here the first term represents the Bragg reflection, and the second represents the scattering proportional to the square of the amplitude of the thermal vibrations, which have a wave vector $\mathbf{k} - \mathbf{K}$.

At the melting point, we may use the classical energy distribution over the vibrational degrees of freedom and then

$$\overline{\left| y_{\mathbf{k}-\mathbf{K}} \right|^2} = \frac{k_B T}{M \nu_{\mathbf{k}-\mathbf{K}}^2}, \tag{62.4}$$

where M is the mass of an ion, and $\nu_{\mathbf{k}-\mathbf{K}}$ is the frequency of the vibrations.

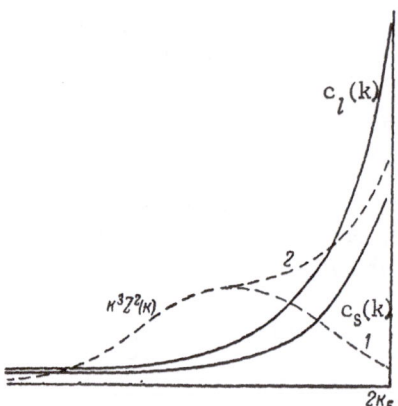

Fig. 30. Correlation function c(k) for a liquid and solid metal, together with the factor $k^3 Z^2(k)$. 1) Plasma scattering; 2) scattering on a pseudopotential.

If **k** lies in the first Brillouin zone, only the limiting waves, whose
velocity is governed by the compressibility β_s, are important. Thus, we obtain

$$c_s\,(0) = \frac{k_B T N \beta_s}{\Omega}, \tag{62.5}$$

which is identical with the expression (61.3) for a liquid.

The formula (62.5) is valid for small values of k, until k approaches the
Brillouin zone surface. Here we enter the region of a thermally broadened
δ peak given by the first term of Eq. (62.3), and $c_s(k)$ should rise rapidly.
Moreover, since k is not parallel to $k - K$, the transverse waves make some
contribution. They have a lower propagation velocity than the longitudinal
waves and therefore they also increase $c_s(k)$.

According to Fig. 30, $c_l(k)$ increases with k faster than does $c_s(k)$ and
therefore the maximum of c(k) for a liquid is broader. To use Eqs. (61.9)
and (61.11), it is necessary to know the behavior of the factor $|Z(k)|^2 k^3$
which occurs in the integral (61.11). This factor may be assumed to be the same
for liquid and solid metals, but according to Fig. 29 it is different for the
plasma scattering and the scattering on the pseudopotential.

Figure 30 gives the expected behavior of the quantities $c_s(k)$, $c_l(k)$, and $|Z(k)|^2 k^3$. It follows from this figure that in the case of sodium, in which the plasma scattering dominates, the main contribution to the integral (61.11) is made by the region of small values of k; according to (61.3) and (62.5), in the first approximation we have

$$\frac{\sigma_s}{\sigma_l} \simeq \frac{c_l(0)}{c_s(0)} = \frac{\beta_l}{\beta_s}. \tag{62.6}$$

For sodium, the ratio of the compressibilities is 1.3, while the ratio of the electrical conductivities is 1.45; the agreement between these two ratios is satisfactory and the difference is easily explained by the fact that $c_l(k)$ rises somewhat faster than does $c_s(k)$.

In other univalent metals, the dominant process is the scattering on the pseudopotential, and the main contribution to the integral (61.11) comes from the region of large values of k, where $c_l(k)/c_s(k) > c_l(0)/c_s(0)$, and the ratio of the conductivities σ_s/σ_l should be higher than that given by Eq. (62.6). This is in agreement with the experimental data presented in Table 1.

Thus Ziman's theory gives a qualitative explanation of the change in the conductivity of various univalent metals on melting. A quantitative theory would require the knowledge of $c(k)$ and $Z^2(k)$ for each of the metals separately. Mott's early theory, presented in § 1, is more quantitative since it includes parameters which are known for some metals.

§ 63. Dependence of the Electrical Conductivity on Temperature and Pressure

In metals, the electron density and the Fermi level, i.e., the average thermal velocity of the electrons taking part in conduction, may be regarded as approximately independent of temperature since the temperature dependence of the electrical conductivity is mainly due to the temperature dependence of the mean free path. According to Eq. (61.9), this dependence has its origin in $c(k)$, because $Z(k)$ is a function only of the ion potential and electron density.

At low values of k, $c_l(k)$ is proportional to the absolute temperature, which follows from Eq. (61.3). Consequently, in liquid sodium, in which the main contribution to the scattering comes from the region of low values of k, the resistivity $1/\sigma_l$ should be approximately proportional to the absolute temperature, which is in agreement with experiment.

In those metals in which the scattering on the pseudopotential is dominant, the temperature dependence of $c_l(k)$ is considerable near $k = 2k_F$. The theory does not give this dependence but the experimental observations indicate that $c_l(k)$ rises only slowly with temperature [9] so that the temperature dependence of the resistivity should be similar. According to the experimental data quoted in [1], $1/\sigma_l$ at constant volume is proportional to T only for sodium and potassium.

For other univalent metals, one term in the resistivity is constant while the other term rises proportionally to T. This is in full agreement with Eq. (48.19), found by a completely different theoretical approach. Ziman's theory gives only an estimate of the coefficient f_1, f_2, f_3, and f_4 of Eq. (48.19) for sodium.

Let us consider the dependence of the electrical conductivity on pressure or volume. If the plasma scattering dominates, this dependence is determined mainly by the derivative $\partial c(0)/\partial \Omega$, which in its turn is governed by the derivative of the compressibility with respect to volume. In liquid and solid metals, the compressibility decreases with decreasing volume; therefore, in the case of sodium, the volume coefficient of the resistivity should be quite large and positive, which is in agreement with experiment.

The structural scattering case is more complex. The effects of pressure on the pseudopotential and on $c(k)$ may partly compensate one another. The total effect may be judged from the fact that the compression of a liquid tends to sharpen the x-ray diffraction maxima, so that near $2k_F$, $c(k)$ may rise and $\partial c(k)/\partial \Omega$ will be negative. Therefore, we may expect the volume coefficient of the resistivity to be small and possibly negative.

§64. Applicability of the Theory to Multivalent Metals [2]

The theory of the present chapter has been applied to univalent metals. Its application to divalent and trivalent metals meets with several obstacles.

The first difficulty is the validity of the weak electron binding approximation in the case of multivalent metals whose pseudopotential is large and cannot be regarded as a small perturbation. In the absence of direct proof, Bardley et al. [2] have argued in favor of the use of the weakly bound electron method in the case of liquid divalent and trivalent metals. These arguments are as follows.

1. The values of the Hall coefficient R_H, measured for several divalent or trivalent metals, are close to the values calculated for free electrons from the formula

$$R_H = \frac{\Omega}{\nu N e},$$

(64.1)

where ν is the number of valence electrons per atom. At the same time, the values of R_H for the same metals differ considerably in the solid state.

2. According to the almost free electron model, the conductivity is

$$\sigma = \frac{4e^2}{3\hbar^2} \tau k^2 \frac{\partial E}{\partial k}.$$

(64.2)

We can calculate $k^2(\partial E / \partial k)$ by measuring σ and determining the transit time τ from the optical data. It is found that for liquid metals of various valences this quantity is close to $\hbar^2 k^3 / m$ and, consequently, the density of states in a liquid metal is the same as for free electrons.

However, the data on the paramagnetic susceptibility and particularly on the Knight shift do not support the free electron model [2]. It seems that the arguments quoted in favor of this model simply confirm the applicability of the effective mass method but not of the almost free electron method, the validity of which seems doubtful in the case of multivalent liquid metals.

The second difficulty is that there are no reliable independent estimates of the value of U for multivalent metals. Therefore, the quantity U^2 has been found by making the calculated resistivities of liquid metals agree with the experimentally measured values. However, for this purpose it is necessary to know the exact value of $\overline{c(k)}$. The value of $c(k)$ has been found by x-ray and neutron diffraction for almost all metals which are of interest and then the expression (60.11) has been calculated by numerical integration. It has been found that $\overline{c}$ is close to unity for the majority of multivalent metals.

The values of U determined from the resistivity have been found to be far too small (about 1 eV), while the forbidden band widths indicate that they should be of the order of several electron volts. Bardley et al. [2] found no satisfactory explanation for this discrepancy. Their suggestion that U(k) is considerably smaller for a liquid than for a solid does not seem to be convincing. Moreover, the whole theory is based on the scattering of electrons by separate ions having the pseudopotential $U(r - R_n)$. If this pseudopotential depends on the mutual distribution of the ions, we cannot consider scattering by separate ions even if we allow for the correlation in the ion distribution.

It follows that the applicability of the theory to multivalent metals is more than suspect.

It is worth noting that the theory developed in the present chapter uses only those quantities which can easily be found from independent experiments: atomic volume, forbidden band width, experimentally determined distribution function, etc. The theory is capable of explaining certain observations, mainly related to the conductivity of univalent metals. On the other hand, the theory needs further development. It has too few mathematical derivations and too many qualitative and not always convincing conclusions. It is difficult to say how promising is the further development of this approach to liquid metals, but it is likely that this theory has a narrower range of applications than those suggested by its authors.

Similar treatments have appeared in the work of Gerstenkorn [10], and of Krishnan and Bhatia [11], but Ziman's theory as presented here is more refined and gives better agreement with experiment. Therefore, we shall not consider the other work [10, 11].

Chapter XII

QUASI-CLASSICAL THEORY
OF AMORPHOUS FERROMAGNETS [1]

§ 65.   Fraction of Antiparallel Spin Pairs

In the preceding chapters, we dwelt on the properties of amorphous and liquid substances connected mainly with the motion of the electrons (i.e., conduction) and with the electron energy spectrum. In the present chapter, we shall consider briefly another quantum-electron property of amorphous substances—ferromagnetism—which is not directly related to the motion of the electrons or their energy spectrum. Therefore, it is simpler to deal with ferromagnetism of amorphous substances than with the properties discussed in earlier chapters. The theory of ferromagnetism of crystals can be easily extended to amorphous substances.

All the known theoretical and experimental studies of ferromagnetism deal with crystals. However, ferromagnetism is due to exchange interactions mainly between the nearest atoms, and strict periodicity in the distribution of atoms does not seem essential. Obviously, ferromagnetism is not observed in liquids because the melting points of ferromagnets are above the Curie point. However, amorphous or glassy substances can exist at very low temperatures so that if they include suitable atoms they may exhibit ferromagnetism.

Amorphous glassy ferromagnets would have advantages in some applications and therefore the search for such ferromagnets is important both from the scientific and practical point of view.

Amorphous and glassy substances usually consist of many components and contain atoms of various types. We shall assume, however, that only atoms of one type are responsible for ferromagnetism, i.e., the exchange integrals between these atoms are positive and large compared with the exchange integrals between other atoms or between ferromagnetic and nonferromagnetic atoms. In the first approximation, such a system may be considered to contain atoms of one type only, the other atoms being regarded as a neutral medium which serves solely to fix average distances between ferromagnetic atoms. From now on, we shall use the term "atoms" to denote "ferromagnetic atoms".

We shall investigate theoretically the conditions for ferromagnetism in an amorphous substance assuming that each atom does not have nearest neighbors at definite equal distances, but that the distances from its neighbors have a random distribution. The calculations are carried out using the "quasi-chemical" method. This method has been developed by Fowler and Gugenheim [2] to deal with alloys and has been applied to ferromagnetism by Stil'bans [3] and Vonsovskii and Shur [4], and we shall follow their treatment.

If N is the total number of atoms, and g(r)dr is the number of neighbors of a given atom which lie at distances from r to r+dr, then the total number of atom pairs at these distances is $1/2$ [Ng(r)dr]. Assume that p(r)dr pairs have antiparallel spins, p^+(r)dr have parallel right-handed spins, and p^-(r)dr have left-handed spins. If A(r) is the exchange integral for atoms separated by a distance r, then the law of mass action for the pairs of atoms with parallel and antiparallel spins is [4]:

$$\frac{p^+(r)\,p^-(r)}{p^2(r)} = \frac{1}{4}\,e^{\frac{4A(r)}{k_B T}}.$$

(65.1)

Vonsovskii and Shur [Eq. (25.18) in [4]] omitted the factor $\frac{1}{4}$ in the above formula. This factor is obtained taking into account the symmetry, regarding a pair with antiparallel spins to be an asymmetrical molecule [Eq. (5.75) in [5], which has a misprint because the fraction $\sigma^2_{ND}/\sigma_{N_2}\sigma_{D_2}$ should be inverted; Stil'bans [3] quotes the above formula correctly].

We shall introduce an average relative magnetization y and a fraction of antiparallel spins x(r). The fraction of right-handed spins is $(1+y)/2$, the fraction of left-handed ones is $(1-y)/2$, and it is easily shown that

$$p(r) = \frac{1}{2}\,Ng(r)\,x(r); \quad p^+(r) = \frac{1}{4}\,Ng(r)[1+y-x(r)];$$

$$p^-(r) = \frac{1}{4}\,Ng(r)[1-y-x(r)].$$

(65.2)

Substituting Eq. (65.2) into (65.1), we obtain a quadratic equation in x:

$$x^2\left(e^{\frac{4A}{k_B T}} - 1\right) + 2x - (1-y^2) = 0.$$

(65.3)

Solving this equation, we find the thermodynamic equilibrium average $\bar{x}$, which is a function of r, the distance between atoms whose electrons form a pair:

$$\bar{x}(r) = \frac{\sqrt{1 + (1 - y^2)\left(e^{\frac{4A}{k_B T}} - 1\right)} - 1}{e^{\frac{4A}{k_B T}} - 1} =$$

$$= \frac{1 - y^2}{\sqrt{1 + (1 - y^2)\left(e^{\frac{4A}{k_B T}} - 1\right)} + 1} \, . \qquad (65.4)$$

§ 66. Average Magnetization and the Curie Point

In order to find the average magnetization y, it is necessary to calculate the free energy of a system F, and to determine its minimum value.

The equilibrium value of that part of the exchange energy of the system which depends on the direction of the spins is

$$E = N \int_0^\infty g(r)\, \bar{x}(r)\, A(r)\, dr = N \int_0^\infty \frac{g(r)\, A(r)\, (1 - y^2)\, dr}{\sqrt{1 + (1 - y^2)\left(e^{\frac{4A}{k_B T}} - 1\right)} + 1} \, .$$

$$(66.1)$$

The free energy may be found from the formula

$$F = T \int_0^{\frac{1}{T}} E\, d\frac{1}{T} - k_B T \ln P, \qquad (66.2)$$

where P is the statistical weight, equal to the number of ways of distributing spins over lattice sites for fixed y:

$$P(y) = \frac{N!}{\left(N\frac{1+y}{2}\right)! \left(N\frac{1-y}{2}\right)!} \, . \qquad (66.3)$$

Vonsovskii and Shur [4] give the free energy formula incorrectly: the right-hand term in their formula should not be squared. Stil'bans [3] uses the formula $F = E - kT \ln P$ instead of Eq. (66.2). This is equivalent to the assumption that E is independent of T for fixed y, which contradicts Eq. (66.1). Therefore, the final formulas obtained by Stil'bans are unreliable.

Substituting Eqs. (66.1) and (66.3) into Eq. (66.2), using Stirling's formula for factorials and assuming that g is independent of T (which is valid for an amorphous substance but not for a liquid), we obtain

$$\frac{2F(y)}{Nk_B T} = (1+y)\ln(1+y) + (1-y)\ln(1-y) - 2\ln 2 +$$

$$+ \frac{1}{2}\int_0^\infty \int_0^{\frac{4A}{k_B T}} \frac{1 - y^2}{\sqrt{1 + (1 - y^2)\left(e^{\frac{4A}{k_B T}} - 1\right) + 1}}\, d\left(\frac{4A}{k_B T}\right) g(r)\,dr.$$

(66.4)

Introducing, for brevity, the notation

$$\sqrt{1 + (1 - y^2)\left(e^{\frac{4A}{k_B T}} - 1\right)} = \alpha(y, r)$$ (66.5)

and carrying out the inner integration in Eq. (66.4), we find

$$\frac{2F(y)}{Nk_B T} = (1+y)\ln(1+y) + (1-y)\ln(1-y) - 2\ln 2 +$$

$$+ \frac{1}{2}\int_0^\infty g(r)\left[(1+y)\ln\frac{\alpha+y}{1+y} + (1-y)\ln\frac{\alpha-y}{1-y} - 2\ln\frac{\alpha+1}{2}\right]dr.$$

(66.6)

From the condition of minimum F ($dF/dy = 0$), we obtain an equation for determining the magnetization y as a function of temperature

$$\ln\frac{1+y}{1-y} + \frac{1}{2}\int_0^\infty g(r)\ln\frac{(\alpha+y)(1-y)}{(1+y)(\alpha-y)}\, dr = 0.$$ (66.7)

This equation always has the solution $y = 0$, but below the Curie point θ this solution represents a free-energy maximum and not a minimum. A free-energy minimum at $T < \theta$ corresponds to $y \neq 0$, i.e., to the presence of spontaneous magnetization.

As is known [4], the Curie point θ is found from the conditions

$$y = 0, \quad \frac{\partial F}{\partial y} = 0, \quad \frac{\partial^2 F}{\partial y^2} = 0. \tag{66.8}$$

The second of these conditions — Eq. (66.7) — is satisfied identically at $y = 0$, and the third gives

$$2 - \int_0^\infty g(r)\left(1 - \frac{1}{\alpha_0}\right) dr = 0; \tag{66.9}$$

here

$$\alpha_0 = \alpha_{y=0} = e^{\frac{2A}{k_B T}}. \tag{66.10}$$

If we assume that $g(r) = z_1 \delta(r - r_1)$, where z_1 is the number of the nearest neighbors, and r_1 is the distance from the nearest neighbor, the formulas obtained here reduce to the formulas for crystals in the nearest-neighbor approximation.

In our calculations, we have not assumed periodicity or rigorous order in the distribution of the atoms. The formulas (66.7) and (66.9) deduced for an amorphous substance do not differ in principle from the corresponding formulas for a crystal. The quantitative difference is due to the dependence of the exchange integral A on the distance r between the atoms. It is known that with increase of r, the integral A tends to become positive. In an amorphous substance, the average distances between atoms are larger than in a crystal with the same short-range order, and therefore some substances, which are not ferromagnetic in the crystalline state, may be ferromagnetic in the amorphous state. On the other hand, at large r the absolute value of A decreases with increase of r; in ferromagnetic crystals, this may reduce y and lower the Curie point on transition to the amorphous state. The scatter of the distances between the atoms and of the values of A may be important if the first maximum of the function $g(r)$ lies near a node of A so that A is negative for some atom pairs. This would considerably lower the Curie point.

Thus, the theory predicts the existence of amorphous ferromagnets but to calculate theoretically the Curie point it is necessary to know $g(r)$ and $A(r)$ for an actual substance.

The formulas can be used for crystals in the second, third, etc., nearest-neighbor approximations if we assume that

$$g(r) = z_1 \delta(r - r_1) + z_2 \delta(r - r_2) + \ldots, \qquad (66.11)$$

where r_1 is the distance to the first nearest neighbors, r_2 to the second nearest neighbors, etc.; z_1, z_2, etc., are the numbers of these neighbors.

The case of a one-dimensional crystal (polymer) is interesting. In this case, $z_1 = 2$, $z_2 = 2$, etc. The substitution of Eqs. (66.11) and (66.10) into Eq. (66.9) gives

$$e^{-\frac{2A_1}{k_B T}} + e^{-\frac{2A_2}{k_B T}} = 1. \qquad (66.12)$$

This equation gives $\theta \neq 0$. Consequently, the well-known result that $\theta = 0$ for $z = 2$, i.e., that a one-dimensional crystal cannot be ferromagnetic, applies only to the nearest-neighbor approximation and not if the exchange interaction between the next neighbors is included. Analysis of Eq. (66.12) shows that even if $A_2 = 0.01 A_1$, the Curie point is only 5 times lower than for $A_2 = A_1$ (four nearest neighbors).

Appendix

THE ELECTRON STATE DENSITY
IN THE BAND "TAIL"
OF AMORPHOUS SEMICONDUCTORS*

The electron state density near a band edge in an amorphous semiconductor was calculated in two ways: by the method of Green's functions, and from the number of local fluctuation levels per unit energy interval.

One of the present authors [1] developed a band theory of amorphous and liquid semiconductors by assuming a small departure, ε, from short-range order in the absence of long-range order. By introducing a deformed coordinate system, the perturbation theory could be used with ε as a small parameter. It was shown that the band structure of the spectrum was retained, accompanied by a slight broadening of the allowed bands and a spreading of their edges. These investigations, however, did not consider the detailed form of the level density $\rho(E)$. Many calculations [2] have shown that in a crystal or a liquid $\rho(E)$ is only perceptibly different near a band edge. However, since the state density in this region considerably influences semiconductor electrical properties, a detailed study of it would be of practical significance.

The aim of the present work, therefore, is to derive an expression for the density of the electron states in the band tail, i.e., in the region where the band edge becomes diffuse. The relative effects of the departures from short-range and long-range order are also investigated.

Two cases will be considered.

---

*By F. M. Gashimzade and A. I. Gubanov, A. F. Ioffee Physico-Technical Institute, Academy of Sciences, USSR, Leningrad. Reprinted from Soviet Physics — Solid State Vol. 6, No. 4, pp. 795-797; translated from Fizika Tverdogo Tela Vol. 6, No. 4, pp. 1030-1033.

1. The neighboring cells are slightly distorted in such a way that the sum of their relative deformations over any region of the crystal is less than unity and the long-range order is retained. This corresponds to a disordered crystalline alloy rather than an amorphous body.

2. The distortions vary slightly from cell to cell, the short-range order being retained in certain regions but differing considerably for widely separated cells and the long-range order being absent. This corresponds more to an inhomogeneously deformed crystal than to a true liquid.

The general perturbation theory can be used for the first case. The Green's function averaged over the deformations of all the cells is given by

$$\langle G^{-1} \rangle = G_0^{-1} - \langle \Sigma \rangle, \tag{1}$$

where

$$\Sigma = \varepsilon^2 \int W_{kk'} G_0(k') W_{k'k} dk' + \varepsilon^3 \int \int W_{kk'} G_0(k') \times$$

$$\times W_{k'k''} G_0(k'') W_{k''k} dk' dk'' + \ldots, \tag{2}$$

W and w are the perturbation operators introduced in [1].

$$G_0^{-1} = E - \frac{k^2}{m^*} - \langle \varepsilon w \rangle - i0, \tag{3}$$

i.e., in Eq. (1) the mean value of the perturbation operator $\varepsilon^2 w$ is included in the zeroth-order Green function. Using the conditions applying to the summation of the deformations of all the cells and taking ε to be small, we obtain

$$\langle G^{-1} \rangle = \langle G_0^{-1} \rangle - \varepsilon^2 A' \int \frac{1}{E - E(\mathbf{k})} \frac{d^3k}{(2\pi)^3} - i\varepsilon^2 \pi A' \int \delta\left(E - E(\mathbf{k})\right) \frac{d^3k}{(2\pi)^3} =$$

$$= \langle G_0^{-1} \rangle - \varepsilon^2 B - i\varepsilon^2 \Gamma. \tag{4}$$

A calculation shows that the ratio of the damping term $\varepsilon^2 \Gamma$ to the intrinsic energy correction term $B\varepsilon^2$ is of the order $\sqrt{E/E_1}$ where E_1 is the energy corresponding approximately to the center of an allowed band. The damping is small for small E (near a band edge). The perturbation theory therefore predicts a shift of the edge by an amount proportional to ε^2. The mean value of the perturbation operator in Eq. (3) only renormalizes the mass.

A spreading of the band edge is therefore not predicted by the perturbation theory when the long-range order is retained, in agreement with the numerous calculations of [2].

Let us now consider the second case, using the method which V. L. Bonch-Bruevich applied to another problem [3]. In a similar way, the averaged Green's function (averaging in an analogous manner to [1]) is given by

$$G(\mathbf{k},\, E) = i \int_0^\infty ds e^{\, i \left(E - \frac{k^2}{m^*} \right) s - A \varepsilon^2 k^4 s^2 - 0 s}, \tag{5}$$

where A is some constant which differs in the one- and three-dimensional cases.

In order to calculate the state density, we must know the imaginary part of the Green's function

$$\rho(E) = \frac{2}{\pi} \int \operatorname{Im} G(\mathbf{k},\, E)\, d\mathbf{k}. \tag{6}$$

From Eq. (5), we obtain after integration

$$\operatorname{Im} G(\mathbf{k},\, E) = \sqrt{\frac{\pi}{4A\varepsilon^2 k^4}}\, \exp\left[-\frac{\left(E - \frac{k^2}{m^*} \right)^2}{4A\varepsilon^2 k^4} \right]. \tag{7}$$

The region of interest is that in which E is negative (E being measured from the bottom of the conduction band of the crystal). It is found that the major contribution to the integral (6) is made by the lower part of the band. Therefore, we restrict ourselves to the integration with respect to k between the limits 0 and $\varkappa$, where $\varkappa$ corresponds approximately to the center of an allowed band, and we can use the effective mass method. As a result, we obtain

$$\rho(E) = \frac{1}{\sqrt{\pi A \varepsilon^2}}\, e^{-\frac{1}{4Am^*\varepsilon^2}} \int_{k < \varkappa} \frac{d\mathbf{k}}{k^2}\, e^{-\frac{|E|}{2Am^*\varepsilon^2 k^2} - \frac{E^2}{4A\varepsilon^2 k^4}}. \tag{8}$$

The asymptote of this expression can be considered either in one or in three dimensions. Here, we shall deal with the three-dimensional case and introduce the variable

$$t = \frac{|E|}{2Am^*\varepsilon^2 k^2},$$

then

$$\rho(E) = \frac{1}{\sqrt{2\pi}} \frac{\sqrt{|E|}}{2A\varepsilon^2 \sqrt{m^*}} e^{-\frac{1}{4Am^{*2}\varepsilon^2}} \int\limits_{U}^{\infty} e^{-Am^{*2}\varepsilon^2 t^2 - t} t^{-3/2} dt, \qquad (9)$$

where

$$U = \frac{|E|}{2Am^*\varepsilon^2 \varkappa^2}.$$

The above expression can be written in terms of Whittaker functions

$$\rho(E) = \frac{1}{\sqrt{2\pi}} \frac{\sqrt{|E|}}{2A\varepsilon^2 \sqrt{m^*}} e^{-\frac{1}{4A\varepsilon^2 m^{*2}}} U^{-3/4} e^{-U/2} \times$$

$$\times \sum_{n=0}^{\infty} \frac{(-A\varepsilon^2 m^* U)^n}{n!} W_{n-3/4;\ n-1/4}(U). \qquad (10)$$

From the known asymptotic expression for Whittaker functions [4], we obtain

$$\rho(E) = \frac{A^{1/2} m^* \varkappa^3 \varepsilon}{\pi^{1/2} |E|} e^{-\frac{1}{4Am^{*2}\varepsilon^2}} e^{-\frac{|E|}{2A\varepsilon^2 m^* \varkappa^2} - \frac{E^2}{4A\varepsilon^2 \varkappa^4}}. \qquad (11)$$

Thus in an amorphous substance the density of states away from the bottom of a "crystalline" conduction band is exponential and the "tail" has a Gaussian distribution with a dispersion $2A^{1/2} \varkappa^2 \varepsilon$. Thus the spreading of the band edge is more important than the edge shift which is proportional to ε^2.

The state density function in the band "tail" can be derived by another method, which clearly illustrates its physical nature. The important role of local fluctuation levels was discussed in [5]. The number of such levels was found to be

$$N = N_0 \int\limits_{b}^{\infty} f(t)\, dt. \qquad (12)$$

Here $f(t) = \frac{1}{\varepsilon \sqrt{2\pi Z}} e^{-\frac{t^2}{2\varepsilon^2 Z}}$, Z is the coordination number, and t the sum of the relative fluctuations of the linear dimensions of the cells. Relationships between the quantities t, the fluctuating potential, and the electron energies were derived; using these relationships (Eqs. (6) and (4) of [5]], we obtain

$$f(x) = \frac{1}{\varepsilon\sqrt{2\pi Z}}\, e^{-\frac{B^2}{2\varepsilon^2 Z}\frac{1}{\left(\sqrt{x}\,\text{arc tg}\,\frac{1}{\sqrt{x}} - 1\right)^2}}, \tag{13}$$

where $x = -E/E_1$ (E_1 is at the center of an allowed band) and B is some constant. For small x, we have

$$f(x) = \frac{1}{\varepsilon\sqrt{2\pi Z}}\, e^{-\frac{B^2}{8\varepsilon^2 Z}}\, e^{-\frac{B^2\pi}{16\varepsilon^2 Z}\sqrt{x}}, \tag{14}$$

and for large x

$$f(x) = \frac{1}{\varepsilon\sqrt{2\pi Z}}\, e^{-\frac{B^2 x^2}{18\varepsilon^2 Z}}. \tag{15}$$

Equation (15) also predicts an exponential "tail" for the density of states in the forbidden band, but, in contrast to Eq. (11), does not contain a linear term in E in the exponent. This is apparently due to the different method of calculation. The similarity of the formulas (11) and (15) confirms that the band "tail" in amorphous semiconductors is due to local fluctuation levels. As was shown in [5], such levels can arise either from a strong deformation of a single cell, or more likely from deformations of the same sign of a group of neighboring cells. The same assumption about the similarity of the deformations of neighboring cells (slightly varying from cell to cell) was made in deriving Eq. (11).

However, in order to use the general perturbation theory, we derived Eq. (4) neglecting both the strong distortions and the similarity of the deformations of neighboring cells, due to the assumed correlation between the deformations. As a result, no band tail was obtained.

Makinson and Roberts [6] have also shown, for a one-dimensional liquid model, that the levels forbidden in a crystal are due to local electron states, the localization being caused by deformations of one sign of a linear chain of cells.

LITERATURE CITED

CHAPTER I

1. N. P. Mokrovskii and A. R. Regel', Zh. Tekhn. Fiz. 23: 2121 (1953).
2. A. Roll and H. Motz, Z. Metallk. 48: 272 (1957).
3. K. Bornemann and G. Ronschenplatt, Metallurgie 9: 473, 505 (1912).
4. E. F. Northrup, J. Franklin Inst. 175: 153 (1913); 177: 1, 287 (1914); 178: 85 (1914).
5. H. Tsutsumi, Sci. Rept. Sendai 7: 93 (1918).
6. Y. Matuyama, Sci. Rept. Sendai 16: 447 (1927).
7. C. C. Bidwell, Phys. Rev. 23: 357 (1924).
8. P. W. Bridgman, Proc. Am. Acad. Arts Sci. 56: 29 (1921); 60: 385 (1925).
9. L. Hackspill, Compt. Rend. 151: 305 (1910).
10. N. S. Kurnakov and A. I. Nikitinski, Z. Anorg. Allgem. Chem. 88: 151 (1914).
11. A. Gutz and N. Boniewsky, J. Chem. Phys. 7: 464 (1909).
12. G. Dodd, Proc. Phys. Soc. (London) B63: 662 (1950).
13. W. Jäger and H. Steinkvehr, Ann. Physik 45: 1089 (1914).
14. H. Kammerlingh-Onnes, Koninkl. Ned. Akad. Wetenschap. Proc. 4: 113 (1911).
15. R. W. Keys, Phys. Rev. 84: 367 (1951).
16. G. Gribe and H. Speidel, Z. Elektrochem. 46: 233 (1940).
17. H. Perlitz, Phil. Mag. 7: 1148 (1926); Acta et commentationes universitatis tartuensis A13 (1928).
18. N. F. Mott, Proc. Roy. Soc. (London) A146: 465 (1934).
19. A. H. Wilson, The Theory of Metals [Russian translation], GITTL, Moscow-Leningrad, 1941.
20. N. Cusack and J. E. Enderby, Proc. Phys. Soc. (London) 75: 395 (1960).
21. G. Busch and O. Vogt, Helv. Phys. Acta 27: 241 (1954).
22. A. R. Regel', Ukr. Fiz. Zh. 7: 833 (1962).
23. N. Cusack and P. W. Kendall, Proc. Phys. Soc. (London) 72: 898 (1958).
24. G. Busch and Y. Tièche, Helv. Phys. Acta 35: 273 (1962).

25. C. A. Kraus and E. W. Johnson J. Phys. Chem. 32: 1289 (1928).
26. G. Borelius and K. Gullberg, Ark. Mat. Astron. och Fysik 31A (17): 1 (1945); G. Borelius, F. Philstrand, I. Anderson, and K. Gullberg, Ark. Mat. Astron. och Fysik 30A (14): 1 (1944).
27. H. Henkels, J. Appl. Phys. 21: 725 (1950).
28. A. I. Blyum and A. R. Regel', Zh. Tekhn. Fiz. 21: 316 (1951).
29. A. I. Blyum, N. P. Mokrovskii, and A. R. Regel', Zh. Tekhn. Fiz., 21: 273 (1951); Izv. Akad. Nauk SSSR, Ser. Fiz. 16: 139 (1952).
30. N. P. Mokrovskii and A. R. Regel', Zh. Tekhn. Fiz. 22: 1281 (1952).
31. N. P. Mokrovskii and A. R. Regel', Zh. Tekhn. Fiz. 23: 779 (1953).
32. A. R. Regel', Structure and Physical Properties of Matter in the Liquid State, Izd. Kievsk. gos. univ. im. Shevchenko, 1954, pp. 117-131.
33. A. R. Regel', Theoretical Problems and Studies of Semiconductors and Semiconductor Metallurgy Processes, Izd. Akad. Nauk SSSR, Moscow-Leningrad, 1955, pp. 12-19.
34. A. F. Ioffe and A. R. Regel', Progr. in Semiconductors (London) 4: 237 (1960).
35. V. M. Glazov and S. N. Chizhevskaya, Fiz. Tverd. Tela 3: 2694 (1961).
36. A. S. Epstein, H. Fritzsche, and K. Lark-Horovitz, Phys. Rev. 107: 412 (1957).
37. V. L. Zyazev and O. A. Esin, Zh. Tekhn. Fiz. 28: 18 (1958).
38. H. Hendus, Z. Naturforsch. 2: 505 (1947).
39. N. P. Mokrovskii and A. R. Regel', Zh. Tekhn. Fiz. 22: 1281 (1952).
40. N. P. Mokrovskii and A. R. Regel', Zh. Tekhn. Fiz. 23: 964 (1953).
41. V. I. Blyum and A. R. Regel', Zh. Tekhn. Fiz. 23: 783 (1953).
42. V. I. Blyum and A. R. Regel', Zh. Tekhn. Fiz. 23: 964 (1953).
43. M. S. Ablova, O. D. Elpat'evskaya, and A. R. Regel', Zh. Tekhn. Fiz. 26: 1366 (1956).
44. F. Gaibullaev and A. R. Regel', Zh. Tekhn. Fiz. 27: 2240 (1957).
45. F. Gaibullaev and A. R. Regel', Zh. Tekhn. Fiz. 27: 1996 (1957).
46. B. T. Kolomiets and N. A. Goryunova, Zh. Tekhn. Fiz. 25: 984 (1955).
47. N. A. Goryunova and B. T. Kolomiets, Zh. Tekhn. Fiz. 25: 2669 (1955).
48. N. A. Goryunova and B. T. Kolomiets, Izv. Akad. Nauk SSSR, Ser. Fiz. 20: 1496 (1956).
49. N. A. Goryunova, B. T. Kolomiets, and V. P. Shilo, Zh. Tekhn. Fiz. 28: 981 (1958).
50. N. A. Goryunova, B. T. Kolomiets, and V. P. Shilo, Fiz. Tverd. Tela 2: 280 (1960).
51. N. A. Goryunova and B. T. Komomiets, Zh. Tekhn. Fiz. 28: 1922 (1958).
52. B. T. Kolomiets and V. P. Pozdnev, Fiz. Tverd. Tela 2: 28 (1960).

53. D. N. Nasledov and E. Malyshev, Zh. Tekhn. Fiz. 16:1127 (1946).

54. M. K. Shidlovskii, Zh. Tekhn. Fiz. 24:837 (1954).

55. W. E. Spear, Proc. Phys. Soc. (London) B70: 669 (1957).

56. N. D. Konozenko, Usp. Fiz. Nauk 52:561 (1954).

57. E. Taft and Z. Apker, Phys. Rev. 96:1456 (1954).

58. Yin Shih-tuan and A. R. Regel', Fiz. Tverd. Tela 3:3614 (1961).

59. T. N. Vengel' and B. T. Kolomiets, Zh. Tekhn. Fiz. 27:2485 (1957).

60. B. T. Kolomiets and T. F. Nazarova, Fiz. Tverd. Tela 2:174 (1960).

61. L. A. Grechanik, N. V. Petrovykh, and V. G. Karpechenko, Fiz. Tverd. Tela 2:2131 (1960).

62. V. A. Ioffe, I. V. Patrina, and S. V. Poberovskaya, Electrical prop - erties of some oxide semiconductor glasses, The Structure of Glass, Vol. 2, Consultants Bureau, New York, 1960, p. 407.

63. M. I. Kornfel'd and L. S. Sochava, Fiz. Tverd. Tela 1:1366, 1370 (1959).

64. M. A. Gilleo, J. Chem. Phys. 19:1921 (1951).

65. E. Billig, Proc. Phys. Soc. (London) B65:216 (1952).

66. B. T. Kolomiets and T. N. Mamontova, Fiz. Tverd. Tela 1:29 (1959).

67. B. T. Kolomiets and B. V. Pavlov, Photoelectric and Optical Phenom- ena in Semiconductors, Kiev, 1959, p. 201.

68. B. T. Kolomiets and V. M. Lyubin, Fiz. Tverd. Tela 2:52 (1960).

69. E. Vogt, Z. Elektrochem. 45:597 (1939).

70. E. Huster, Ann. Physik 33 (5):477 (1938).

71. M. F. Deigen, Zh. Eksperim. i Teor. Fiz. 26:293, 300 (1954).

CHAPTER II

1. D. R. Hartree, Proc. Cambridge Phil. Soc. 24:89 (1928).

2. V. Fock, Zh. Physik 61:126 (1930).

3. F. Seitz, Modern Theory of Solids [Russian translation] GITTL, Mos- cow-Leningrad, 1949.

4. G. G. Gel'man, Quantum Chemistry, ONTI, Moscow-Leningrad, 1937.

5. L. A. Schmid, Phys. Rev. 92:1373 (1953).

6. S. P. Shubin and S. V. Vonsovskii, Sov. Fiz. 7:292 (1935); 10:348 (1936).

7. S. V. Vonsovskii, Tr. Inst. Fiz. Metal., Akad. Nauk SSSR, Ural'sk. Fillial 12:9 (1949).

8. S. V. Vonsovskii and E. N. Agafonova, Collection on the Seventieth Birthday of Academician A. F. Ioffe, Izd. Akad. Nauk SSSR, 1950, p. 92.

9. S. V. Vonsovskii, Usp. Fiz. Nauk 48: 289 (1952).

10. S. V. Vonsovskii and B. V. Paduchev, Zh. Eksperim. i Teor. Fiz. 25: 510 (1953).

11. S. V. Vonsovskii and V. S. Galishev, Zh. Eksperim. i Teor. Fiz. 25: 584 (1953).

12. S. V. Vonsovskii and Yu. M. Seidov, Izv. Akad. Nauk SSSR, Ser. Fiz. 18: 319 (1954).

13. S. V. Vonsovskii, Zh. Tekhn. Fiz. 25: 2022 (1955).

14. N. N. Bogolyubov and S. V. Tyablikov, Zh. Eksperim. i Teor. Fiz. 19: 251, 256 (1949).

15. A. G. Samoilovich and K. G. Tovstyuk, Zh. Tekhn. Fiz. 27: 1753 (1957).

16. A. I. Gubanov and A. A. Nran'yan, Fiz. Tverd. Tela 1: 1044 (1959).

17. D. Bohm and D. Pines, Phys. Rev. 82: 625 (1951); 85: 338 (1952); 92: 609 (1953).

18. D. Pines, Phys. Rev. 92: 626 (1954).

19. D. Pines, Solid State Phys. 1: 367 (1955).

20. P. Debye and E. Hückel, Physik. Z. 24: 185 (1923).

21. F. Bloch, Z. Physik. 52: 555 (1928).

22. J. C. Slater, Phys. Rev. 87: 807 (1952).

23. J. R. Reitz, Solid State Phys. 1: 1 (1955).

24. F. C. von der Lage and H. A. Bethe, Phys. Rev. 71: 612 (1947).

25. G. G. Hall, Phil. Mag. 43: 338 (1952).

26. A. G. Samoilovich and K. D. Tovstyuk, Zh. Tekhn. Fiz. 27: 1753 (1957).

27. A. A. Nran'yan, Fiz. Tverd. Tela 2: 474, 1650 (1960).

28. J. Bardeen, Phys. Rev. 52: 688 (1937).

29. H. A. Bethe and A. Sommerfeld, Electron Theory of Metals [Russian translation], ONTI, Moscow-Leningrad, 1938.

30. H. Fröhlich, Proc. Roy. Soc. (London) 166: 230 (1937).

31. B. M. Davydov and I. M. Shmushkevich, Usp. Fiz. Nauk 24: 23 (1940).

32. F. J. Blatt, Solid State Phys. 4: 199 (1957).

33. W. Kohn, Solid State Phys. 5: 257 (1957).

34. S. I. Pekar, Investigation of the Electron Theory of Crystals, Moscow-Leningrad, 1951.

35. C. Kittel and A. Mitchell, Phys. Rev. 96: 1488 (1954).

36. W. Kohn and J. M. Luttinger, Phys. Rev. 97: 1721 (1955); 98: 915 (1955).

37. D. L. Dexter, Phys. Rev. 93: 244 (1954).

38. J. A. Krumhansl, Phys. Rev. 93: 245 (1954).

39. M. F. Deigen, Zh. Eksperim. i Teor. Fiz. 33: 773 (1957).

40. B. S. Gourary and F. J. Adrian, Phys. Rev. 105: 1180 (1957); 106: 1356 (1957); 107: 488 (1957).

41. I. V. Aborenkov, Vestn. Leningr. Univ. 22: 14 (1958).

CHAPTER III

1. Ya. I. Frenkel', Kinetic Theory of Liquids, Izd. Akad. Nauk SSSR, Moscow-Leningrad, 1945.
2. A. I. Gubanov, Zh. Tekhn. Fiz. 17:525 (1947).
3. J. A. Prins, Naturwissenschaften 19:435 (1931).
4. B. E. Warren, J. Appl. Phys. 8:645 (1937).
5. V. I. Danilov, Scattering of X Rays in Liquids, ONTI, Moscow-Leningrad, 1935.
6. I. V. Radchenko, Usp. Fiz. Nauk 61:249 (1957).
7. P. Debye and H. Menke, Physik. Z. 31:797 (1930).
8. V. I. Danilov and V. E. Neimark, Zh. Eksperim. i Teor. Fiz. 5:724 (1935).
9. J. Campbell and J. Hildebrand, J. Chem. Phys. 11:330 (1943).
10. P. C. Sharrah and G. P. Smith, J. Chem. Phys. 21:228 (1953).
11. H. Hendus, Z. Naturforsch. 2a (9):505 (1947).
12. A. E. Glauberman and V. P. Tsvetkov Dokl. Akad. Nauk SSSR 106:623 (1956).
13. A. E. Glauberman, Zh. Eksperim. i Teor. Fiz. 23:249 (1952).
14. J. D. Bernal and R. H. Fowler, J. Chem. Phys. 1:515 (1933).
15. G. Stewart, Rev. Mod. Phys. 2:116 (1930).
16. B. E. Warren, Phys. Rev. 44:969 (1933).
17. Structure of Glass, Proc. Conference on Structure of Glass, Izd. Akad. Nauk SSSR, Moscow-Leningrad, 1955.
18. W. H. Zachariasen, J. Am. Ceram. Soc. 54:3841 (1932).
19. A. A. Lebedev, Izv. Akad. Nauk SSSR, Ser. Fiz. 4:584 (1940).

CHAPTER IV

1. A. I. Gubanov, Zh. Eksperim. i Teor. Fiz. 26:139 (1954).
2. G. P. Boev, Theory of Probability, GITTL, Moscow-Leningrad, 1950.
3. H. A. Bethe and A. Sommerfeld, Electron Theory of Metals [Russian translation] ONTI, Moscow-Leningrad, 1938.
4. Ya. I. Frenkel', Wave Mechanics, Gostekhizdat, Moscow-Leningrad, 1937.
5. A. I. Gubanov, Fiz. Tverd. Tela 2:651 (1960).
6. A. I. Gubanov, Fiz. Tverd. Tela 3:2164 (1961).
7. A. I. Gubanov Fiz. Tverd. Tela 4:1510 (1962).
8. R. Landauer and J. C. Helland, J. Chem. Phys. 22:1655 (1954).

9. R. E. B. Makinson and A. P. Roberts, Australian. J. Phys. 13:437 (1960).
10. M. Lax and J. C. Phillips, Phys. Rev. 110:41 (1958).
11. K. K. Rebane, Tr. Inst. Fiziki i Astron., Akad. Nauk Est. SSR, No. 5: 72 (1957).
12. R. Eisenschitz and P. Dean, Proc. Phys. Soc. (London) 70:713 (1957).

CHAPTER V

1. A. I. Gubanov, Zh. Eksperim. i Teor. Fiz. 28:401 (1955).
2. L. D. Landau and I. M. Lifshits, Quantum Mechanics, GITTL, Moscow, 1948.
3. A. I. Gubanov, Zh. Tekhn. Fiz. 27:2510 (1957).
4. F. Hund and B. Mrowka, Ber. der Sachs. Akad. der Wissensch., Math. Phys. 87:185, 325 (1935); F. Hund, Physik. Z. 36:888 (1935).

CHAPTER VI

1. A. I. Gubanov, Zh. Eksperim. i Teor. Fiz. 30:862 (1956).
2. S. Shubin, Zh. Eksperim. i Teor. Fiz. 3:461 (1933).
3. H. A. Bethe and A. Sommerfeld, Electron Theory of Metals [Russian translation] ONTI, Moscow-Leningrad, 1938.
4. F. Seitz, Modern Theory of Solids [Russian translation], GITTL, Moscow-Leningrad, 1949.
5. A. I. Gubanov, Zh. Tekhn. Fiz. 26:1651 (1956).
6. I. Z. Fisher, Dokl. Akad. Nauk SSSR 117:39 (1957).

CHAPTER VII

1. A. I. Gubanov, Zh. Eksperim. i Teor. Fiz. 31:462 (1956).
2. H. A. Bethe and A. Sommerfeld, Electron Theory of Metals [Russian translation], ONTI, Moscow-Leningrad, 1938.
3. Ya. I. Frenkel', Kinetic Theory of Liquids, Izd. Akad. Nauk SSSR, Moscow-Leningrad, 1945.
4. I. M. Ryzhik, Tables of Integrals, Sums, and Products, GITTL, Moscow-Leningrad, 1948.
5. H. Fröhlich, Proc. Roy. Soc. (London) 160:230 (1937).
6. B. I. Davydov and I. M. Shmushkevich, Usp. Fiz. Nauk 24:23 (1940).

CHAPTER VIII

1. A. I. Gubanov, Zh. Tekhn. Fiz. 27:3 (1957).
2. L. Brillouin, Quantum Statistics [Russian translation], Kiev, 1934.
3. A. E. Glauberman, Some Problems in the Kinetic Theory of Systems of Interacting Particles, author's abstract of dissertation for Doctorate of Physicomathematical Sciences, Akad. Nauk SSSR, Leningrad, 1956.

CHAPTER IX

1. A. P. Roberts and R. E. B. Makinson, Proc. Phys. Soc. (London) 79:630 (1962).
2. A. I. Gubanov, Fiz. Tverd. Tela 3:2336 (1961).
3. I. M. Lifshits, Zh. Eksperim. i Teor. Fiz. 17:1017 (1947).
4. G. T. Koster and J. C. Slater, Phys. Rev. 95:1167 (1954).
5. Ya. Kouteskii and A. Fingerland, Dokl. Akad. Nauk SSSR 125:841 (1959).
6. A. I. Gubanov, Fiz. Tverd. Tela 4:2873 (1962).
7. G. G. Gel'man, Quantum Chemistry, ONTI, Moscow-Leningrad, 1937.
8. A. I. Gubanov and A. A. Nran'yan, Fiz. Tverd. Tela 1:1044 (1959).
9. J. Bardeen and W. Shockley, Phys. Rev. 80:72 (1950).
10. Ya. I. Frenkel', Kinetic Theory of Liquids, Izd. Akad. Nauk SSSR, Moscow-Leningrad, 1945.
11. I. Z. Fisher, Fiz. Tverd. Tela 1:192 (1959).
12. H. Reiss, J. Chem. Phys. 25:681 (1956).
13. F. E. Williams, J. Chem. Phys. 19:457 (1951).
14. N. D. Potekhina, Opt. i Spektroskopiya 2:388 (1957).
15. A. I. Gubanov, Zh. Eksperim. i Teor. Fiz. 16:523 (1946).

CHAPTER X

1. S. F. Edwards, Phil. Mag. 6:617 (1961).
2. S. F. Edwards, Proc. Roy. Soc. (London) 267:518 (1962).
3. V. L. Bonch-Bruevich and S. V. Tyablikov, Method of Green's Functions in Statistical Mechanics, Fizmatgiz, Moscow, 1961.
4. A. A. Abrikosov, L. P. Gor'kov, and L. P. Dzyaloshinskii, Quantum Field Theory Methods in Statistical Physics, Fizmatgiz, Moscow, 1962.
5. W. Kohn and J. M. Luttinger, Phys. Rev. 108:590 (1957).

6. S. F. Edwards, Phil. Mag. 3:1020 (1958); Russian translation in collection: Problems of the Quantum Theory of Irreversible Processes, V. L. Bonch-Bruevich, (ed.), Moscow, 1962.
7. V. Heine, The Electronic Band Structure of Liquid Metals (a preprint), Cavendish Laboratory, Cambridge.
8. W. D. Knight, Solid State Phys. 2:93 (1956).

CHAPTER XI

1. J. M. Ziman, Phil. Mag. 6:1013 (1961).
2. C. C. Bardley, T. E. Faber, E. G. Wilson, and J. M. Ziman, Phil. Mag. 7:865 (1962).
3. N. H. Cohen and V. Heine, Advan. Phys. 7:399 (1958); Phys. Rev. 122:1821 (1961).
4. J. M. Ziman, Advan. Phys. 10:1 (1961).
5. J. G. Collins and J. M. Ziman, Proc. Roy. Soc. (London) 264:60 (1961).
6. J. C. Phillips and Z. Kleinman, Phys. Rev. 116:287 (1959).
7. N. Cusack and J. E. Enderby, Proc. Phys. Soc. (London) 75:395 (1960).
8. G. Fournet, Handbuch der Physik 32:238 (1958).
9. N. S. Gingrich and L. Heaton, J. Chem. Phys. 34:873 (1961).
10. N. Gerstenkorn, Ann. Physik. 10:49 (1952).
11. K. S. Krishnan and A. B. Bhatia, Nature (London) 156:503 (1948).

CHAPTER XII

1. A. I. Gubanov, Fiz.Tverd. Tela 2:502 (1960).
2. R. H. Fowler and E. Gugenheim, Proc. Roy. Soc. (London) A174:189 (1940).
3. L. S. Stil'bans, Zh. Eksperim. i Teor. Fiz. 9:432 (1939).
4. S. V. Vonsovskii and Ya. S. Shur, Ferromagnetism, GITTL, Moscow-Leningrad, 1948.
5. R. H. Fowler and E. Gugenheim, Statistical Thermodynamics [Russian translation], IL, Moscow, 1949.

APPENDIX

1. A. I. Gubanov, Zh. Eksperim. i Teor. Fiz. 26: 139 (1954); 28: 401 (1955).
2. R. Makinson and A. Roberts, Australian J. Phys. 13: 437 (1960).
3. V. L. Bonch-Bruevich, Fiz. Tverd. Tela 4: 2660 (1952).
4. I. M. Ryzhik and I. S. Gradshtein, Tables of Integrals, Sums, Series, and Products, Third ed., GITTL, Moscow, 1961, p. 430.
5. A. I. Gubanov, Fiz. Tverd. Tela 4: 2873 (1962).
6. R. Makinson and A. Roberts, Proc. Phys. Soc. (London) 79: 630 (1961).

Publisher's Note

The following Russian journals cited in this book
are available in cover-to-cover translation

Russian title	Title of translation (translation began)	Publisher
Doklady Akademii Nauk SSSR	Soviet Physics— Doklady (1956)	American Institute of Physics
Fizika Tverdogo Tela	Soviet Physics— Solid State (1959)	American Institute of Physics
Izvestiya Akademii Nauk SSSR: Seriya Fizicheskaya	Bulletin of the Academy of Sciences of the USSR: Physical Series (1954)	Columbia Technical Translations
Optika i Spektroskopiya	Optics and Spectroscopy (1959)	American Institute of Physics
Uspekhi Fizicheskikh Nauk	Soviet Physics— Uspekhi (1958)	American Institute of Physics
Zhurnal Eksperimental'- noi i Teoreticheskoi Fiziki	Soviet Physics— JETP (1955)	American Institute of Physics
Zhurnal Tekhnicheskoi Fiziki	Soviet Physics— Technical Physics (1956)	American Institute of Physics

INDEX